Energy Production, Load and Battery Management Framework with Supporting Methods for Smart Microgrids

Francky Catthoor • Dafang Zhao •
Ittetsu Taniguchi • Pavlos S. Georgilakis •
Dimitrios Soudris • Kostas Siozios •
Andreas Kazantzidis
Editors

Energy Production, Load and Battery Management Framework with Supporting Methods for Smart Microgrids

Springer

Editors
Francky Catthoor
IMEC
Leuven, Belgium

Dafang Zhao
Osaka University
Osaka, Japan

Ittetsu Taniguchi
Osaka University
Osaka, Japan

Pavlos S. Georgilakis
Zografou Campus
National Technical University of Athens
Athens, Greece

Dimitrios Soudris
Zografou Campus
National Technical University of Athens
Athens, Greece

Kostas Siozios
Aristotle University of Thessaloniki
Thessaloniki, Greece

Andreas Kazantzidis
University of Patras
Patras, Greece

ISBN 978-3-031-92024-0 ISBN 978-3-031-92025-7 (eBook)
https://doi.org/10.1007/978-3-031-92025-7

This work was supported by IMEC, University of Patras, Osaka University and National Technical University of Athens.

This Springer imprint is published by the registered company Springer Nature Switzerland AG
The registered company address is: Gewerbestrasse 11, 6330 Cham, Switzerland

Preface

Distributed, localized energy grids and also smart grids are an important emerging topic in the energy domain, especially to support the growing amount of renewable energy production sources. Today, most of these are organized in large solar or wind parks, but the domestic market share is rapidly growing too. Without such more distributed and even smart grids, the traditional regional energy grids will not be able to cope with the growing renewable content.

In this book we provide an overview of the results of a broad university collaboration addressing this emerging topic, including also the related work comparison. To our knowledge, today no comprehensive book focuses on this content, and we believe now is the time to fill that gap.

In addition, we have many promising and quite consolidated results from our consortium research activities from the past decade. These have been published as individual contributions in point papers but never with the embedding in the overall context and goals of our collaborative project. So this book is the first public way to disseminate that overall holistic vision. It also shows how this work is situated in the existing related work, and how it differentiates from that.

Several unique aspects are present in our research methodologies and approach. The short-term wind and PV forecasting techniques are based on physics-based modeling, which has allowed us to obtain unprecedented accuracy levels within reasonable run-times, compatible with online monitoring and forecasting as needed for smart grids.

The HVAC load modeling and forecasting is based on a fully up-scalable approach including aggregation techniques and with the focus on obtaining high accuracy without requiring much (typically not available) local information on the HVAC installation.

The overall energy management framework combining renewable production, HVAC loads, and battery storage is based on multiple time-scales. This unique approach has enabled to arrive at near-Pareto-optimal working points, with again reasonable run-times on embedded processing platforms.

The book is organized in five chapters.

Chapter 1 provides the overall vision and the holistic approach. It also puts our work in a global context.

Chapter 2 focuses on PV energy yield modeling specifically for distributed localized energy grids. PV energy yield modeling is conducted with respect to the special characteristics, constraints, and limitations of distributed localized energy grids. It addresses short-term PV energy yield forecasting (minutes to hours ahead). It highlights the most important features of short-term forecasting horizons (e.g., fine-grained resolution, sky images, global optimization).

Chapter 3 does the same for wind turbine energy yield forecasting.

Chapter 4 introduces a load modeling and management framework that takes into consideration energy/cost availability from renewable sources in order to decide upon future actions (e.g., refine cooling/heating strategy). It evaluates alternative considerations for historical weather data in order to decide upon optimum (in terms of energy consumption and thermal comfort) configuration of the HVAC system. It also considers aspects related with the implementation of a control system as part of a low-cost embedded device (deep-edge device).

Chapter 5 provides an overview of our multi-timescale energy management approach. That framework considers the balancing of renewable energy production and consumption by loads. But it also integrates battery storage operation, power purchases from the regular grid, and aggregation of appliance usage. It addresses the challenges of maintaining energy balance and reducing electricity costs in smart PV systems due to the unpredictability and fast variation of renewable energy generation.

Acknowledgments We would like to acknowledge the contribution of the many colleagues who have contributed to the research we have summarized in this book. We deeply appreciative the collaboration in the past with the renewable energy research groups at IMEC, Leuven and Genk, Belgium; at Osaka University, Japan; at NTUAthens, Greece; and at EnergyVille and KULeuven, Belgium.

We extend our heartfelt thanks to the past cooperation with all colleagues in our collaboration network, especially with Hans Goverde, Patrizio Manganiello, Eli Shirazi, Lieve Helsen, Johan Driesen, Geert Deconinck, Jef Poortmans, Toshihiro Suzuki, Yoshiyuki Shimoda, Kenshiro Kato, and Koki Iwabuchi.

The research in this book is partly supported by the Japan Society for the Promotion of Science (21J10312, 22H03697, 24K20901), Daikin Industries, Ltd., next to our own organizations.

October 2024 The book and chapter authors

The original version of the book has been revised. A correction to this book can be found at https://doi.org/10.1007/978-3-031-92025-7_6.

Contents

Acronyms

ANN	Artificial Neural Network
AS	Appliance Scheduling
CC	Cloud Class
CGEM	Coarse-Grained Energy Management
CMM	Cloud Motion Modeling
CMV	Cloud Motion Vector
CRPS	Continuous Ranked Probability Score
CWC	Coverage Width Criterion
DRED	Dutch Residential Energy Dataset
ELM	Extreme Learning Machine
Eyield	Energy yield
EMS	Energy Management System
FGEM	Fine-Grained Energy Management
GSI	Ground-based Sky Image
HEMS	Home Energy Management System
HVAC	Heating, Ventilation, and Air Conditioning
LSTM	Long Short-Term Memory
LUBE	Lower Upper Bound Estimation
MCKP	Multiple Choice Knapsack Problem
MCT	Multi-Color Threshold
MIP	Mixed Integer Programming
MINLP	Mixed-Integer Non-Linear Programming
MPC	Model Predictive Control
MSE	Mean Squared Error
NARX	Non-linear Autoregressive Network with eXogenous inputs
NLP	Non-Linear Programming
NWP	Numerical Weather Prediction
PID	Proportional-Integral-Derivative
PI	Prediction Interval
PICP	Prediction Interval Coverage Probability
PINRW	Prediction Interval Normalized Root Width

PPD	Predicted Percentage of Dissatisfied
PV	Photovoltaic
PVPP	Photovoltaic Power Plant
RF	Random Forest
RMSE	Root Mean Squared Error
SAMURAI	Simulated Annealing Module with Update Range of moves, Adaptive annealing schedule, and Inner-loop criterion
SOC	State-Of-Charge
TOU	Time-Of-Use
WF	Wind Farm
WPD	Wavelet Packet Decomposition
WPPF	Wind Power Probabilistic Forecasting
WT	Wind Turbine

Chapter 1
Motivation, Focus, and Contributions of This Book

Francky Catthoor, Ittetsu Taniguchi, Dafang Zhao, Kostas Siozios, Pavlos S. Georgilakis, and Andreas Kazantzidis

1.1 Smart Energy Systems

With the increasing emphasis on and growth of environmental awareness in society, a large amount of renewable energy such as solar and wind power have been introduced to demand-side consumers to reduce CO_2 emissions and electricity costs [16]. However, renewable energy is intermittent and uncontrollable, and its generation fluctuates in the short term due to various environmental factors, especially due to strong variation in wind speed, cloud movement, and local temperature. This fluctuation nature will inevitably cause supply-demand mismatches [9]. To address this issue, a smart energy system, which integrates renewable energy production and demand control, including temporary energy storage like batteries, will make such systems energy-efficient, more resilient, and significantly more eco-friendly [10].

F. Catthoor (✉)
Zografou Campus, National Technical University of Athens, Athens, Greece
e-mail: catthoor@microlab.ntua.gr

I. Taniguchi · D. Zhao
University of Osaka, Osaka, Japan

K. Siozios
Aristotle University of Thessaloniki, Thessaloniki, Greece

P. S. Georgilakis
School of Electrical and Computer Engineering, National Technical University of Athens, Athens, Greece

A. Kazantzidis
Laboratory of Atmospheric Physics, Physics Department, University of Patras, Patras, Greece

F. Catthoor et al. (eds.), *Energy Production, Load and Battery Management Framework with Supporting Methods for Smart Microgrids*,
https://doi.org/10.1007/978-3-031-92025-7_1

1.2 Energy Management Systems

Several key technologies have to be hence combined in these smart energy systems. The main one is a global energy management system (EMS) to manage the balance of energy in real time among renewable generation, appliances, battery systems, and electricity purchases [4]. The EMS aims to improve several criteria, including minimizing electricity costs, reducing demand peaks, and maximizing the self-sufficiency rate of renewable sources [11]. A major requirement is to be run-time efficient, as we do not want to execute this on cloud servers due to the necessity to provide online real-time control. And the cost for the power plant owner or the local community should be reasonable so compute platforms like the well-known Raspberry Pi [14] are more appropriate to use in this context than local servers which are powerful but more costly to purchase and to maintain.

So in order to create this run-time efficiency, we have introduced multiple concurrently applied timescales with dedicated solvers at each of these scales. In real-time energy management, the fluctuation of renewable generation such as wind turbine and photovoltaic (PV) systems causes severe energy loss and imbalances. The time constant involved in those is subsecond. So at the fastest timescale in our EMS framework, we work with a resolution of 1 second to effectively deal with those fluctuations. Concurrently, at slower timescales with a resolution of 15–30 minutes, scheduling distributed energy resources (DERs) such as a battery storage system and a multitude of shiftable appliances is important to shift energy demand from peak to off-peak hours and to reduce electricity costs. To fill the gap between fast and slow timescales in energy management, our framework provides the glue between these scales, and it repeatedly solves three consecutive optimization problems at certain time intervals. This multi-time scale organization significantly reduces computational complexity while maintaining excellent solution quality.

More information on our proposed approach for this is provided in Chap. 5.

1.3 Renewable Energy Yield Modeling and Prediction

However, the unpredictable and short-term fluctuations of renewable sources will negatively impact the performance of the EMS [13]. It is hence of paramount importance to reduce the errors in predicting the energy yield (Eyield) at all the timescales needed within the EMS. That is a daunting challenge because the environmental monitors usually have significant noise related to their operation, and it is very common to have missing sample points during their usage, which adds to the uncertainty. Moreover, the weather models are known to have a large amount of uncertainty when forecasting the future evolution.

In order to reduce the impact of these noise and uncertainty contributions, we have employed several effective counter measures:

1. It is well-known that both for photovoltaic (PV) modules and wind turbines, the errors in longer time periods (at least 1 hour ahead) are strongly averaging out, so the positive and negative impact largely cancel each other. That property has been effectively exploited in the literature on hour(s)-ahead forecast methods [3, 5]. So also in our mid-term Eyield forecasts, we do rely on this heavily; see Chap. 2 for more details on the PV production and Chap. 3 for the wind turbine option. Both of these approaches are based on empirical base models which are optimized exploiting state-of-the-art machine learning techniques.
2. But for the (ultra-)short-term forecasts which are minutes ahead, this averaging out property is not usable yet, so instead we rely on accurate physics-based modeling foundations. It is crucial then to incorporate all relevant dimensions. For the wind turbines, these are mostly the wind field which is heavily impacted by the terrain. For the PV modules, next to the electrical and optical aspects, sufficient care should be taken for modeling the local wind and temperature profiles. For the longer timescales, this would not be feasible because of exploding run-times, but for a few minutes up to (a maximum of) 30 minutes ahead prediction, our earlier research has shown that this is practically achievable even on cheap compute platforms like the earlier mentioned Raspberry Pi. More details on this are again found in Chap. 2 for the PV production and Chap. 3 for the wind turbine option.

   The main advantage of these physics-based models is that they can reliably handle extrapolations because the trends are accurately incorporated. Based on these equations, it is also possible to represent correlations which are present in all these real renewable energy installations. It is well-known that noise sources and other fluctuations are mostly uncorrelated, so by using the proper techniques, it is then even feasible to recover useful information which is close to or even partly hidden below the average noise level! With these powerful techniques, our final accuracy levels are surprisingly high, even on days with strongly variable weather. They do still require a proper initial calibration for the installation in the site at hand. But for that purpose, modern machine learning algorithms are a big help to improve the accuracy of that calibration phase.
3. An additional method to improve the mid-term forecasts is to incorporate probabilistic modeling to provide bounds. Many of these are based on prediction intervals (PIs) and we have further improved those. Also more detailed information on this is provided in the respective chapters on PV and wind turbines.
4. The machine learning techniques used in the state of the art typically rely on gradient-based foundations which are known to be (very) prone to becoming stuck in local optima. So in order to reduce that risk, we have complemented the Eyield predictions with hill-climbing techniques, like simulated annealing, genetic search, and swarm optimizers. These algorithms are traditionally very run-time demanding, but we have started to reduce that overhead, by introducing dynamically adapting versions, in particular for simulated annealing. More detailed information on all this is again provided in the respective chapters on PV and wind turbines.

1.4 Demand-Side Modeling

Another crucial component for the EMS is the modeling and prediction of the demand-side evolution in time. Today, the major contribution for households is what has been termed HVAC, i.e., heat, ventilation, and air-conditioning. But it is expected that also the amount of electrical vehicles (EV) will grow, so for the future it is certainly necessary to provide accurate models for that segment too. The other household appliances like washing machines are typically less demanding when aggregated over long periods.

It is still a challenge to have access to public large databases which have collected realistic EV data over sufficiently long time traces and with a rich variation of usage patterns. Hence, in our approach we have mainly focused on the HVAC case up to now. Here, the aggregation aspect to effectively handle a huge amount of small individual contributions has been a crucial contribution. But we have also put significant effort to develop an approach which is both sufficiently accurate for our context and which is at the same time also very user-friendly. So it relies on very limited information to be provided by the site owners, and it can be considered as mostly a black-box model. This is in contrast to the potentially extremely accurate white-box modeling techniques in the literature which however also require a huge amount of calibration information of the building; see, e.g., [1]. More details on our approach can be found in Chap. 4.

1.5 Temporary Energy Storage

Another major way which is currently employed by the EMS to mitigate the environmental fluctuations is to install energy storage systems like large batteries. These can counter the temporary shortage or surplus of energy by the production side on a fast timescale [12]. However, they are quite costly and not environmentally friendly. Hence, it is crucial to reduce their need as much as possible. For this purpose we have worked on several countermeasures:

1. It is anyway desirable to improve battery modeling and properly incorporate workload-dependent storage in the EMS [2, 15]. But the accuracy of state-of-the-art modeling approaches is either too run-time consuming (for the physics-based options) or too poor (for the empirical models). So we have invested effort in providing a hybrid alternative which combines very good accuracy with acceptable run-time efficiency; see [17]. A summary is provided in Chap. 5.
2. However, an even more effective way would be to introduce knobs at the production side. We have come up with a way to achieve this for PV modules by exploiting their strong temperature dependence. Our PV production knob proposal has been published in [7].

1.6 Dynamic Pricing

Another interesting and promising feature of smart energy systems is that one can potentially influence the demand side needs by so-called dynamic pricing [8]. This topic will not be incorporated in the book material, but we have worked on several approaches which are compatible with our overall EMS framework. More details can be found in [6].

References

1. Atam, E., Helsen, L.: Control-oriented thermal modeling of multizone buildings: Methods and issues: intelligent control of a building system. IEEE Control Syst. Mag. **36**(3), 86–111 (2016). https://doi.org/10.1109/MCS.2016.2535913
2. Azuatalam, D., Paridari, K., Ma, Y., Förstl, M., Chapman, A.C., Verbič, G.: Energy management of small-scale PV-battery systems: a systematic review considering practical implementation, computational requirements, quality of input data and battery degradation. Renew. Sustain. Energy Rev. **112**, 555–570 (2019)
3. Bazionis, I., Kousounadis-Knousen, M., Georgilakis, P., Shirazi, E., Soudris, D., Catthoor, F.: A taxonomy of short-term solar power forecasting: classifications focused on climatic conditions and input data. IET Renew. Power Gener. **17**(9), 2411–2432 (2023). https://doi.org/10.1049/rpg2.12736
4. Chen, C., Duan, S., Cai, T., Liu, B., Hu, G.: Smart energy management system for optimal microgrid economic operation. IET Renew. Power Gener. **5**(3), 258–267 (2011)
5. Giebel, G., Brownsword, R., Kariniotakis, G., Denhard, M., Draxl, C.: The State-Of-The-Art in Short-Term Prediction of Wind Power: A Literature Overview, 2nd edn. ANEMOS.plus (2011). https://doi.org/10.11581/DTU:00000017. Project funded by the European Commission under the 6th Framework Program, Priority 6.1: Sustainable Energy Systems
6. Iwabuchi, K., Kato, K., Watari, D., Taniguchi, I., Catthoor, F., Shirazi, E., Onoye, T.: Flexible electricity price forecasting by switching mother wavelets based on wavelet transform and long short-term memory. Energy AI **10**, 100192 (2022)
7. Iwabuchi, K., Watari, D., Zhao, D., Taniguchi, I., Catthoor, F., Onoye, T.: Enhancing grid stability in PV systems: a novel ramp rate control method utilizing PV cooling technology. Appl. Energy (2024). https://doi.org/10.1016/j.apenergy.2024.124737
8. Khan, A.R., Mahmood, A., Safdar, A., Khan, Z.A., Khan, N.A.: Load forecasting, dynamic pricing and DSM in smart grid: a review. Renew. Sustain. Energy Rev. **54**, 1311–1322 (2016)
9. Liang, X.: Emerging power quality challenges due to integration of renewable energy sources. IEEE Trans. Ind. Appl. **53**(2), 855–866 (2017)
10. Lund, H., Østergaard, P.A., Connolly, D., Mathiesen, B.V.: Smart energy and smart energy systems. Energy **137**, 556–565 (2017)
11. Mariano-Hernández, D., Hernández-Callejo, L., Zorita-Lamadrid, A., Duque-Pérez, O., Santos García, F.: A review of strategies for building energy management system: Model predictive control, demand side management, optimization, and fault detect & diagnosis. J. Build. Eng. **33**, 101692 (2021)
12. Merabet, A., Tawfique Ahmed, K., Ibrahim, H., Beguenane, R., Ghias, A.M.Y.M.: Energy management and control system for laboratory scale microgrid based Wind-PV-Battery. IEEE Trans. Sustain. Energy **8**(1), 145–154 (2017)
13. Olatomiwa, L., Mekhilef, S., Ismail, M.S., Moghavvemi, M.: Energy management strategies in hybrid renewable energy systems: a review. Renew. Sustain. Energy Rev. **62**, 821–835 (2016)

14. Richardson, M., Wallace, S.: Getting Started with Raspberry Pi: Electronic Projects with Python, Scratch, and Linux. Maker Media (2014)
15. Szumanowski, A., Chang, Y.: Battery management system based on battery nonlinear dynamics modeling. IEEE Trans. Vehicular Technol. **57**(3), 1425–1432 (2008)
16. Varun, Prakash, R., Bhat, I.K.: Energy, economics and environmental impacts of renewable energy systems. Renew. Sustain. Energy Rev. **13**(9), 2716–2721 (2009)
17. Zhao, D., Taniguchi, I., Catthoor, F., Onoye, T.: Transient response and non-linear capacity variation aware unified equivalent circuit battery model. In: Proceedings of 8th World Conference on Photovoltaic Energy Conversion, pp. 1579–1584 (2022)

Chapter 2
Accurate PV Energy Yield Forecasting

Markos A. Kousounadis-Knousen, Dimitris Anagnostos, Ioannis K. Bazionis, Apostolos Bakovasilis, Pavlos S. Georgilakis, and Francky Catthoor

In recent years, photovoltaic (PV) installations have recorded the highest yearly growth rates among renewable technologies. In 2023, the global cumulative PV capacity increased by nearly 450 GW, reaching a total of 1.6 TW. Continuous advancements in PV technologies have enhanced their cost-benefit ratio by reducing manufacturing costs while increasing solar cell efficiency. Furthermore, modern PV systems exhibit increased flexibility, which has led to a substantial growth in building-integrated small-scale PV installations.

PV generation is characterized by high variability and volatility, mainly due to the presence and movement of cloud formations. [27]. Due to changes in the sky conditions, PV generation can drop by up to 70% in just a few seconds [6], while fluctuation frequencies can increase up to 1 Hz for PV modules with thinner glass [17]. These fluctuations can have severe effects on the operation of modern microgrids. Therefore, broader integration of PV systems necessitates the introduction of smart energy management systems (EMSs) [35], which in turn rely on accurate forecasts of PV generation.

Optimal EMSs include multiple timescales (see Chap. 5); thus, forecasts of PV generation need to be issued for different forecasting horizons. Figure 2.1 illustrates the horizon-based classification of PV forecasting we propose, including

---

M. A. Kousounadis-Knousen (✉) · I. K. Bazionis · P. S. Georgilakis
School of Electrical and Computer Engineering, National Technical University of Athens, Athens, Greece
e-mail: kousounadisknousen@mail.ntua.gr

D. Anagnostos · F. Catthoor
Zografou Campus, National Technical University of Athens, Athens, Greece

A. Bakovasilis
IMEC-EnergyVille, Genk, Belgium

F. Catthoor et al. (eds.), *Energy Production, Load and Battery Management Framework with Supporting Methods for Smart Microgrids*,
https://doi.org/10.1007/978-3-031-92025-7_2

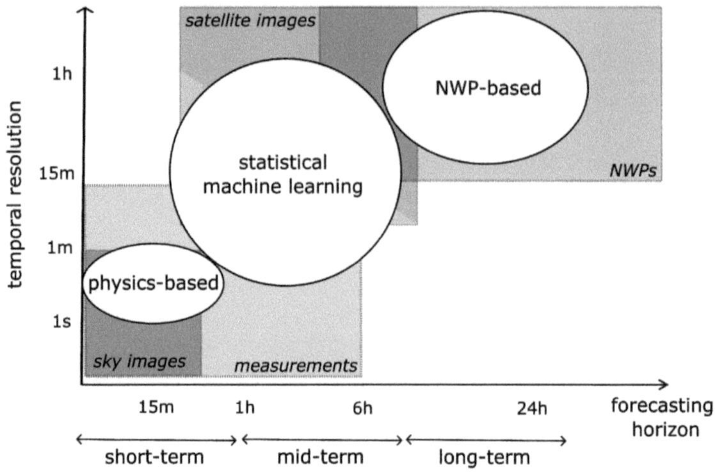

Fig. 2.1 Classification of PV power forecasting based on the forecasting horizon, along with preferred modeling approach, input data type, and output temporal resolution

the preferred input type, modeling approach, and forecasting resolution. Minute-scale forecasts enable specific real-time EMS operations, such as ramp rate control, demand balancing, and battery scheduling [4, 6].

Compared to the prevailing literature, we distinguish ourselves in two ways:

1. Regarding the forecasting horizon, we adopt the classification used in [14]: intra-hour (short-term), intra-day (mid-term), and day-ahead (long-term). Longer forecasting horizons are omitted due to their decreased importance in the scheduling and operation of modern microgrids.
2. Regarding the modeling approach, we consider physics-based models as the optimal choice for minute-scale modeling of PV generation. In such fine-grained minute-scale or second-scale resolutions, the temporal averaging effect is negligible, resulting in increased variability. This is illustrated in Fig. 2.2, in which a daily PV power curve is plotted for three different temporal resolutions. Physics-based models effectively respond to the highly dynamic conditions recorded in minute-scale resolutions. Section 2.1.2 provides more details about this consideration. However, due to their computationally intensive nature, we recommend the usage of physics-based models only for short-term forecasting horizons, for which the total number of forecasts can still be generated in reasonable run-times.

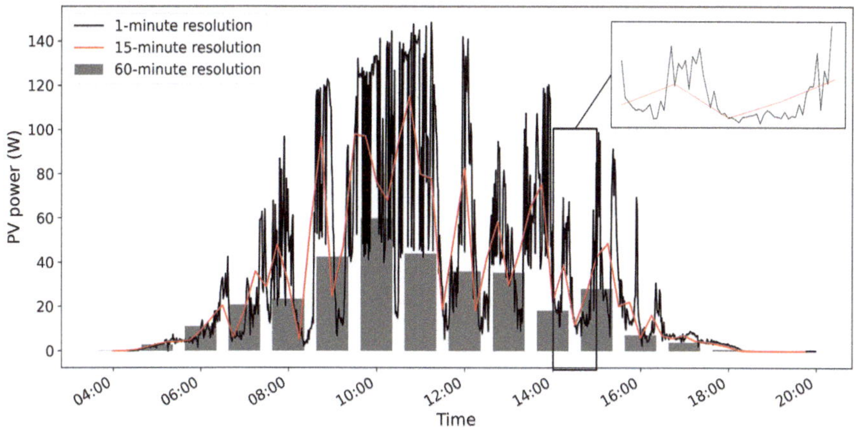

Fig. 2.2 PV power generation of a 180W PV module in three different temporal resolutions, during a cloudy day

2.1 Basic Principles of Short-Term PV Energy Yield Forecasting

Minute-scale short-term PV energy yield (Eyield) forecasting is particularly challenging due to the reduced averaging effect when modeling at fine-grained temporal resolutions. The forecasting challenge is further amplified for small-scale, distribution-level PV systems integrated into buildings and urban areas, for which the spatial smoothing effect is negligible. Efficient short-term PV Eyield forecasting requires careful consideration of the input type, the modeling approach and calibration, as well as the computational constraints. In this section, we briefly present the basic principles of minute-scale short-term PV Eyield forecasting.

2.1.1 Ground-Based Sky Images

The type of input data significantly affects the performance of minute-scale short-term PV Eyield forecasting models. Ground-based Sky Images (GSIs) constitute the most important input type, as they provide a complete picture of the current sky condition at very high spatiotemporal resolutions. Furthermore, valuable insight into short-term cloud movement can be obtained by comparing several consecutive GSIs [7]. GSIs can be captured either by specialized camera setups called All Sky Imagers or by simple fish-eye-lensed surveillance cameras. The choice of the correct camera type depends on the forecasting setup requirements. Further information about different camera types and GSIs can be found in [25].

PV generation can be accurately forecasted in the short term using GSIs as a stand-alone input option, without the need for on-site PV-related numerical measurements. However, we advocate combining GSIs with on-site PV-related numerical measurements to form hybrid input sets that encompass all necessary elements for accurate minute-scale short-term PV power forecasting. Common PV-related numerical measurements include the incident solar irradiance, the ambient and module temperature, the wind speed and direction, as well as PV power itself. Basic preprocessing of GSIs is discussed in Sect. 2.3.

2.1.2 Physics-Based PV Modeling

Forecasting the output of small-scale distribution-level PV systems under fast-varying weather conditions and minute-scale temporal resolutions necessitates detailed models that account for the optical, thermal, and electrical properties of PV cells [2]. Such a physics-based PV model was introduced in [19], and is abstractly presented in Fig. 2.3. Unlike most physics-based models, which typically incorporate only the electrical component of PV cells, the model introduced in [19] effectively couples optical, thermal, and electrical sub-models, to accurately calculate the Eyield of MWT PV modules even under non-steady state conditions and second-scale temporal resolutions. The optical component consists of a layer-by-layer absorption model [19]. The illumination absorption rates of each layer are passed as inputs to the thermal and electrical components. Compared to the optical and thermal components, the electrical component is much easier to model accurately; thus, simple empirical equations based on the two-diode model are adequate.

It is well-known that the efficiency of PV modules is directly affected by cell temperature. However, the thermal state of PV cells is often either overlooked or considered uniform in physics-based PV models. In reality, the cell temperature is a

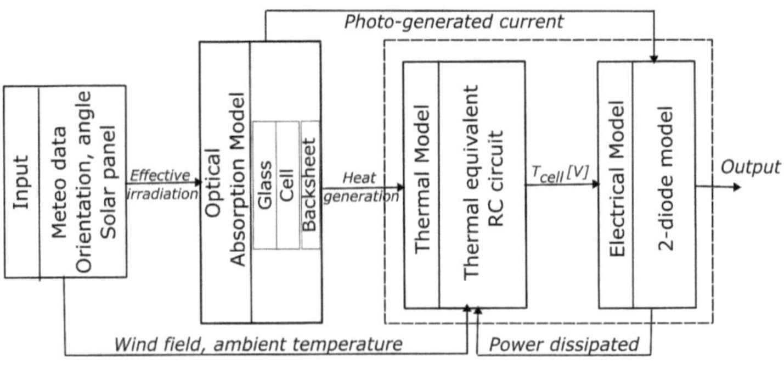

Fig. 2.3 Schematic overview of optical-thermal-electrical single-cell PV model introduced in [19]

result of complicated intra-module heat flows caused by conduction and convection of heat, as well as radiation effects. These heat flows are further affected by the interaction between the PV module and its environment [12]. Studies have shown that heat distribution varies significantly due to wind flows near the PV module's surface [12, 16]. For example, the temperature difference between the top and back surface of PV modules can exceed 20 °C, depending on the wind flows relative to the module's position [18].

The physics-based model introduced in [19] accounts for the complex heat flows by incorporating thermal RC-equivalent sub-circuits, heat transfer coefficients, and circuit coupling. The Stefan-Boltzmann law is used for the radiative heat losses from the module's surface to the environment. Spatially resolved heat transfer coefficients are experimentally extracted through extensive simulations, to model the intra-module and inter-module convection of heat, primarily forced by wind flows. Heat conduction due to semi-illumination of the PV module is addressed by coupling the RC-equivalent sub-circuit of each cell with the sub-circuits of its neighboring cells. We refer to [12, 13, 18–20] for more information regarding the physics-based PV model of Fig. 2.3, as well as the experimental simulations.

In contrast to simplified physics-based models or parametric statistical approaches, the fine-grained physics-based model of Fig. 2.3 retains its accuracy levels regardless of the environmental conditions, with an average error ranging from 3.6% to 5.5% [20]. When forecasting PV Eyield with a temporal resolution of 1 second, the average percentage error remains below 20% during the whole forecasting horizon [3]. The average error drops below 10% and 5% for temporal resolutions of 10 seconds and 1 minute, respectively [4]. Furthermore, the temporal lag between the forecasted PV energy yield curve and the actual output is 5 seconds for a 5-minute forecasting horizon and remains below 3 minutes for forecasting horizons up to 15 minutes [4]. It is thus obvious that a detailed PV model, such as the one shown in Fig. 2.3, can effectively forecast PV Eyield in the short term, even under dynamic weather conditions and ultrahigh spatiotemporal resolutions.

2.1.3 PV Model Calibration

Regardless of the modeling approach, optimal configuration of a PV model plays a pivotal role in minute-scale short-term forecasting. Calibrating the coefficients of a physics-based model, the weights of an artificial neural network, or the hyper-parameters of a statistical model, can be a challenging task that significantly affects the generated outcome. Especially in cases of non-steady state conditions where minute-scale temporal resolutions are required, the shape of the objective function to be minimized during parameter fine-tuning is usually non-convex, anisotropic, and full of local minima.

The optimization algorithm employed for fine-tuning minute-scale PV models should meet three basic search criteria: (i) the algorithm should not be sensitive to the initial search point, (ii) the algorithm should be able to perform some kind of systematic hill climbing to avoid getting stuck in local minima, and (iii) given enough iterations, the algorithm should guarantee convergence. Exhaustive search algorithms meet these criteria; however, they are not preferred due to scalability concerns. On the other hand, gradient descent optimization algorithms are highly sensitive to their initialization and prone to getting stuck in local minima when dealing with non-convex objective functions. In Sect. 2.2, we propose a novel hybrid global optimization algorithm that satisfies the basic search criteria while maintaining the scalability of the overall forecasting approach.

2.1.4 Energy Yield as a Target Variable

Most studies use PV power as the target variable when forecasting the output of PV systems in the short term. Over the course of the forecasting horizon, forecasts are expressed in the form of PV power values, averaged over the specified forecasting interval. However, using Eyield as the target variable in minute-scale forecasting can significantly reduce the average forecasting error. Unlike PV power forecasting models, the bias of which often increase with the forecasting horizon, PV Eyield models maintain relatively stable bias and error levels [3]. Additionally, real Eyield curves are particularly valuable for energy management tasks that require fine-grained resolutions, such as battery scheduling. Thus, henceforth we will refer to PV Eyield instead of PV power when forecasting in the short term.

Forecasting PV Eyield directly, rather than forecasting the power output and then converting it to energy, is also beneficial for minute-scale short-term forecasting. The power accumulation process can cause an additive effect on the forecasting bias, further decreasing the accuracy toward the end of the forecasting horizon. Furthermore, directly forecasting PV Eyield eliminates all extra errors induced by interconnected forecasting models. Chained irradiance-to-power-to-Eyield models introduce additional forecasting errors which are propagated to the final forecasts. For more details on the advantages of direct PV Eyield minute-scale forecasting, we refer to [3].

2.1.5 Computational Limitations

Minute-scale short-term PV Eyield forecasting often involves forecasting intervals of less than 1 minute and forecasting resolutions as high as 1 second [3]. These tight time windows provide little room for computationally intensive models that require long inference times [11]. Additionally, the optimization duration should be limited to allow local online learning and calibration, if necessary [5]. Thus, computational efficiency is crucial for the applicability and operational feasibility of a minute-scale short-term PV Eyield forecasting model.

Another important objective, often neglected in minute-scale short-term PV Eyield forecasting, is to minimize the required computational resources [10]. Minute-scale short-term PV Eyield forecasts are particularly useful for small-scale, low-level residential PV systems, to enable fine-grained tasks related to community-level EMSs. Thus, the cost of the overall framework should be kept within reasonable limits, to maintain affordability for the average consumer. The PV Eyield forecasting model, from the algorithm execution to the processing and storage of the data, should be developed to run on cost-effective hardware platforms, such as the Raspberry Pi [15]. In this direction, it is essential to ensure that the PV Eyield forecasting model can perform satisfactorily with cheap cameras even in the absence of other sensors [36].

2.1.6 Adaptability and Resilience

The continuous integration of small-scale low-level building and urban PV systems poses additional challenges in the development of minute-scale short-term PV Eyield forecasting models. To allow smooth integration, PV models should be robust toward highly varying and dynamic environments, adapting to different types, configurations, and installations of PV modules. Furthermore, the overall forecasting approach should be operable regardless of the surrounding context of the GSIs, the camera settings, the prevailing weather conditions, etc. Developing case-specific short-term PV Eyield forecasting models would be economically and environmentally unfeasible. Thus, adaptability and resilience are essential aspects to be considered during the design of short-term PV Eyield forecasting models.

These highly varying and dynamic environments are the main reasons we avoid the usage of deep leaning for PV modeling, despite the promising results deep learning has showcased in various research fields. Deep learning usually relies on pretrained models and elaborate representative data to adapt to varying and dynamic environments. However, this generalization approach is too coarse-grained for the generation of minute-scale short-term PV Eyield forecasts. Deep learning models struggle to capture the microscopic interactions of PV modules, briefly described in Sect. 2.1.2. Even in absence of a state-of-the-art physics-based PV model, we would still advocate for shallower data-driven models, due to their computational efficiency, increased interpretability, and sufficient nonlinear approximation capabilities, given the appropriate optimization framework.

2.2 Hybrid Global Optimization of the Short-Term PV Energy Yield Model

In this section, we introduce a hybrid optimization algorithm [28] that balances exploration and exploitation by combining an adaptive hill-climbing approach with classic gradient descent optimization. The proposed hybrid optimization algorithm satisfies the search criteria listed in Sect. 2.1.3, with respect to the computational limitations and resilience requirements of minute-scale short-term PV Eyield forecasting. Implementation details of the proposed hybrid optimization algorithm are presented in Sect. 2.4.

2.2.1 Background

Since its inception, simulated annealing [26] has been consistently used to solve optimization problems in various fields, mainly due to its systematic hill-climbing logic and global exploration capabilities. Given a sufficiently large number of iterations, SA can guarantee convergence to a solution close to the global optimum. However, SA is generally considered computationally inefficient, as too many iterations are usually required to ensure a consistent performance. Therefore, SA has seen multiple extensions, particularly at the height of its popularity during the 1980s and the 1990s.

Such a simulated annealing extension, called Simulated Annealing Module with Update Range of moves, Adaptive annealing schedule, and Inner-loop criterion (SAMURAI), was introduced in [9] to improve the computational efficiency of classic SA. The hill-climbing approach of SAMURAI ensures sufficiently good estimates of the global optimum with the fewest possible evaluation steps. Furthermore, all parameters of SAMURAI are fully adaptive and problem independent. Thus, SAMURAI stands as an optimal option for the parameter fine-tuning of a minute-scale short-term PV Eyield forecasting model.

2.2.2 Hierarchical Partitioning of the Objective Function

SAMURAI works particularly well for problems with some form of hidden hierarchy in the objective function [9]. By the term hidden hierarchy, we mean that the objective function can be partitioned into hierarchical levels which are not necessarily macroscopically visible. The most critical decisions during the search process are made in regions that contain transitions to lower hierarchical levels. Therefore, more iterations should be spent in these regions to guarantee that the optimal decision will be made. As the temperature decreases, the resolution of the search becomes higher, to allow the identification of more detailed hierarchical levels.

A simplified schematic example of the SAMURAI search process in a one-dimensional cut of a non-convex objective function E is shown in Fig. 2.4. Each snapshot corresponds to a different level of hierarchy. The horizontal arrows represent the range of the "jumps" performed to explore new states. At higher temperatures, SAMURAI coarsely evaluates the search space to spot the most

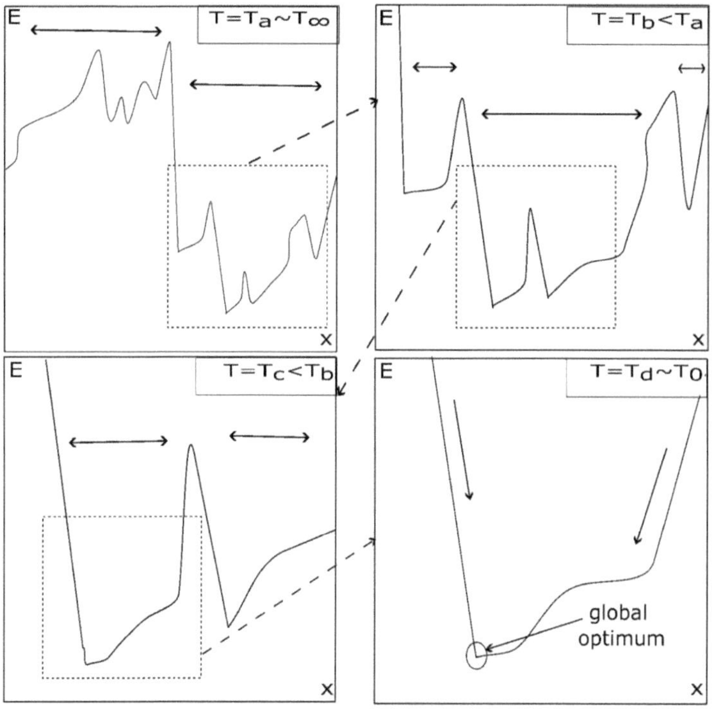

Fig. 2.4 Simplified schematic example of the SAMURAI search process in a one-dimensional space. Each snapshot corresponds to a different level of hierarchy

promising areas. The jumps are big, and the resolution of the search is low. As the temperature decreases, lower levels of hierarchy are reached, the "jumps" become smaller, and less uphill solutions are accepted. At each hierarchy level, the objective function is evaluated at a higher resolution, and the search is directed toward the most promising areas. During the final temperature values, the search is completely localized to allow the identification of the global optimum.

2.2.3 Global Annealing Schedule with SAMURAI

The inner-loop evaluations in SAMURAI are modeled as Markov chains. At each temperature step t, new states are added to the chain until the following equilibrium criterion is satisfied:

$$\frac{1}{N_k^{(t)}} \sum_k e^{\frac{E_{ref}^{(t)} - E_k^{(t)}}{T^{(t)}}} \leqslant \delta \tag{2.1}$$

where $E_k^{(t)}$ is the objective function value of state k, $E_{ref}^{(t)}$ is the objective function reference value at temperature $T^{(t)}$, $N_k^{(t)}$ is the number of states explored up to state k, and δ is a user-specified meta-parameter which controls the level of search granularity at each temperature step. If 2.1 is satisfied, the cumulative contribution of all states is relatively small, indicating that an equilibrium has been reached at temperature $T^{(t)}$ and no new states need to be added to the chain. This way, the Markov chain is longer in critical regions of the search space, to allow comprehensive exploration and ensure the right search direction will be chosen.

The new states are selected based on Monte Carlo analysis with random selection. The move range should be adaptively limited as the temperature decreases, to gradually restrict the initial global search and allow convergence. The type of moves and the move range limitation strategy depend on the problem formulation and the data type. The acceptance of new states is based on the Metropolis criterion, with which uphill solutions are accepted with an exponentially decreasing probability.

The annealing schedule of SAMURAI is determined by the following equation:

$$\alpha^{(t)} = \begin{cases} \alpha_{max} - \left(\alpha_{max} - \alpha^{(t-1)}\right) \frac{M^{(t-1)}}{M^{(t)}} & M^{(t)} > M^{(t-1)} \\ \alpha^{(t-1)} - \left(\alpha^{(t-1)} - \alpha_{min}\right) \frac{M^{(t-1)} - M^{(t)}}{M^{(t-1)} - 2} & M^{(t)} \leqslant M^{(t-1)} \end{cases} \tag{2.2}$$

where $\alpha^{(t)}$ is the temperature ratio between current temperature step t and previous temperature step $t - 1$, α_{min} and α_{max} are user-specified ratio bounds, and $M^{(t)}$ is the total number of explored states (inner-loop iterations) at temperature step t. At each temperature step, variable a changes proportionally to the length difference between the current and the previous Markov chain. The temperature decrease is steeper at less critical regions of the search space to speed up convergence. On the contrary, temperature decreases at a slower pace in critical regions to guarantee that all necessary search paths will be explored.

The initial temperature value is experimentally determined by iteratively updating a random starting value until the average acceptance ratio becomes acceptable [1]. The goal is to initialize the temperature such that all new states are accepted during the first few temperature steps. While this initialization method may not be the optimal choice in terms of computational efficiency, it allows the coarse-grained exploration of the whole search space, which is an essential prerequisite of global searching.

2.2.4 Algorithm Interface and Local Search

When the lowest level of hierarchy is reached (last snapshot of Fig. 2.4), the objective function tends to become monotonic. At this point, SAMURAI struggles to reach the global minimum, and spends too many iterations to converge. Since the neighborhood of the global minimum has already been found, a faster local search algorithm with a better convergence rate is more appropriate. Such an algorithm is the gradient descent-based Adaptive moment estimation (Adam) optimizer.

In the proposed hybrid optimization approach, Adam is connected in series with SAMURAI, to boost local convergence without compromising the global nature of the final solution. An effective global-to-local decision criterion should be selected to reliably determine the interface between SAMURAI and Adam. When the temperature reaches a final value T_0, the annealing stops, and the optimized set of parameters serve as the initial point for re-optimization with Adam. The final temperature value should be determined adaptively, with respect to the corresponding average acceptance probability of uphill solutions. If the average acceptance probability of uphill solutions at temperature $T^{(t)}$ is close to zero, there is no longer a need for SAMURAI, and thus $T^{(t)}$ is considered the final temperature:

$$T^{(t)} = T_0 \quad if \quad \frac{1}{N^{(t)}_{k,uphill}} \sum_{N} e^{\frac{E^{(t)}_{ref} - E_k}{T^{(t)}}} < \beta \tag{2.3}$$

where $N^{(t)}_{k,uphill}$ is the number of uphill solutions explored at temperature step t and β is the user-specified meta-parameter which controls the threshold of the uphill solution average acceptance probability, below which the algorithm can switch to Adam.

2.3 Ground-Based Sky Images Processing

Appropriate processing of GSIs is essential to obtain an accurate picture of the current sky conditions and gain insight on the movement of the clouds. Figure 2.5 illustrates a basic processing example procedure for GSI-based short-term PV Eyield forecasting. GSIs can be converted to numerical data through feature extraction or convolutional neural networks. Common features extracted from GSIs

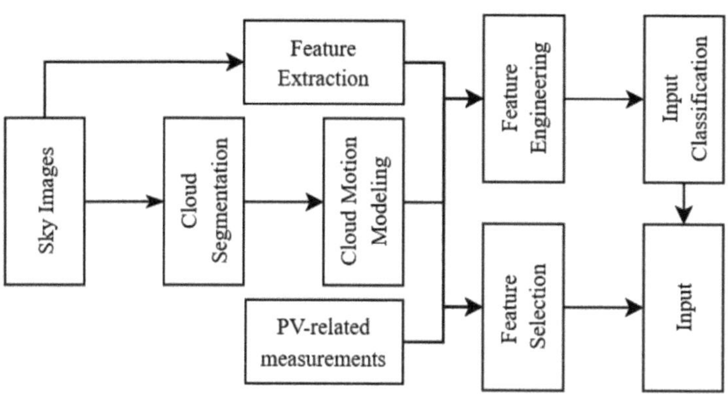

Fig. 2.5 GSI processing block diagram for short-term PV Eyield forecasting

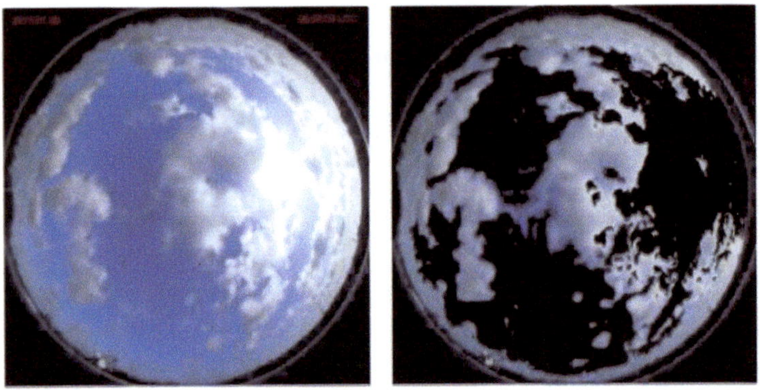

Fig. 2.6 Cloud segmentation results generated with the MCT technique

include spectral features, e.g., the mean Red (R) color, the mean Blue (B) color, the standard deviation of B, the difference between R and B, as well as textural features, e.g., energy, entropy, contrast, and homogeneity [22]. Local features are also often extracted, with methods such as the Scale-Invariant Feature Transform. [33].

Cloud formations are recognized in GSIs with cloud segmentation (CS) techniques. There are several CS techniques available in the literature, each with its own advantages and limitations [21]. The most simple and efficient CS technique is based on the pixelwise difference of R and B [22]. The R-B difference threshold is a good indicator of a cloudy pixel. An improvement of classic threshold-based CS is the Multi-Color Threshold (MCT) technique [24], which introduces two additional thresholds, to reduce the tendency of the R-B difference threshold in underestimating the total cloud cover. An example of the CS results generated with the MCT technique is presented in Fig. 2.6. The biggest limitation of the MCT technique is its sensitivity to threshold selection. The optimal threshold values can dramatically change with different camera setups and calibrations. Furthermore, as seen in Fig. 2.6, the MCT technique struggles around the circumsolar area.

CS techniques are useful in identifying the presence of clouds and eventually estimating the total cloud cover of the sky. On the other hand, cloud movement is estimated with Cloud Motion Modeling (CMM). The objective of CMM is to extract Cloud Motion Vectors (CMVs) by comparing multiple consecutive images. The positions of the clouds are then estimated for each step of the forecasting horizon, to get insight into future sky conditions. CMM is particularly important during periods with frequent switching between cloudy and sunny conditions. In such cases, it is extremely challenging to forecast the energy yield of PV modules without the addition of CMVs to the input [31]. Effective CMM requires the image acquisition frequency to be as close to as possible to 1 Hz. Popular CMM methods include optical flow and block matching [31]. An increasingly popular alternative to CMM is the usage of convolutional neural networks.

The selection of the appropriate input features significantly affects the performance of a PV model. Feature selection for physics-based PV models depends

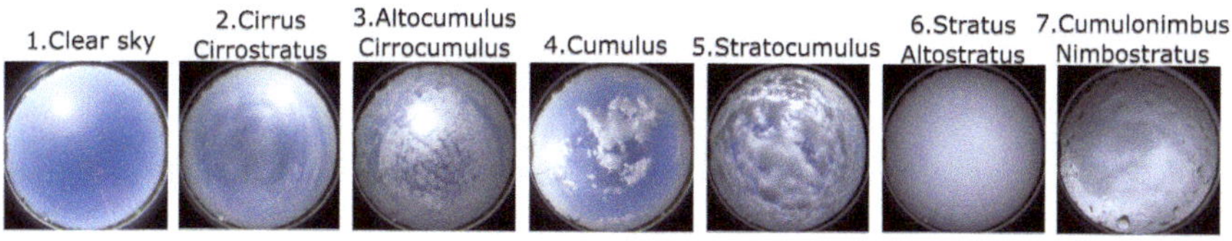

Fig. 2.7 GSI classification based on International Cloud Classification System [22]

on the model's specifications. On the other hand, the input for data-driven models should be carefully chosen to facilitate the approximation of PV power generation. Besides PV-related measurements, relevant input information includes the solar zenith and azimuth angles, the total cloud cover, and the average R-B difference. In the presence of clouds, and regardless of the modeling approach, CMVs should be appropriately processed to generate cloud movement forecasts. The rest of the input features can be kept constant throughout the forecasting horizon, as their effect on the thermal state of PV modules does not drastically change in short timescales [17].

The operation of PV modules significantly differs under different sky conditions; thus, it is essential to fine-tune the PV model independently for different sky conditions to ensure a consistent modeling performance. Classification of GSIs is usually based on cloud types, as defined by the International Cloud Classification System [22]. This approach defines seven distinct classes, shown in Fig. 2.7. The first class corresponds to clear sky conditions. The remaining classes correspond either to a specific cloud type or to a combination of similar cloud types.

Classification of GSIs becomes a challenge in the absence of ground-truth labels. Most GSI datasets do not contain labels, and there is no obvious method to automatically generate semantic labels based on cloud types. Therefore, manual labeling of the GSIs is a prerequisite to train supervised classifiers. However, identifying different types of clouds is a difficult task that requires human expertise. More importantly, manually labeling several hundreds or thousands of images is often not feasible. It is thus necessary to consider alternative classification options, such as clustering. Unlike supervised classification methods, clustering automatically identifies subgroups within datasets, without the need for ground-truth labels. Besides not needing ground-truth labels, unsupervised classification of GSIs has another major advantage: the selection of the classification criteria and the optimal number of clusters can be directed toward optimizing PV Eyield forecasts, by combining clustering with feature engineering techniques and co-optimizing them with the PV Eyield forecasting model. Supervised classification methods, on the other hand, usually classify GSIs based in predetermined criteria that were designed for other tasks, not necessarily related to PV Eyield forecasting. An example of the most representative GSIs of each of the seven clusters generated by applying k-means clustering on an encoded feature dataset is depicted in Fig. 2.8. Note that the number of clusters is purposely chosen equal to the number of classes of Fig. 2.7 to facilitate comparison.

Fig. 2.8 GSI classification generated by k-means clustering, for seven clusters. Each GSI corresponds to the most representative image of a cluster

2.4 Short-Term PV Energy Yield Forecasting Model Implementation

2.4.1 Case Study

The case study consists of a 180W rooftop PV module located at the University of Oldenburg. The forecasting horizon is set to 15 minutes, while the forecasting resolution is kept as high as 1 second to eliminate the temporal averaging effects. The input data used, described in detail in [32], are a combination of meteorological data, PV-related measurements, and GSIs. Measurements are recorded every second and include the diffuse horizontal irradiance, the direct normal irradiance, the temperature, and the power output of the PV module. The GSIs are retrieved every 10s by a low-cost fish-eye-lens camera. Features such as the average of R, G, and B channels, the difference between R and B, standard deviation of B, and image contrast are extracted from the GSIs. CMVs are extracted by the optical flow algorithm. The information about the future sky condition is then provided in the form of RGB forecasts. Further information about the input data retrieval and preprocessing can be found in [32].

The final input instances are a combination of the RGB forecasts and the image features. The classification of the input is based on the classification suggested in [22] (Fig. 2.7). The labels of the GSIs have already been manually preselected in [32]. Even though the dataset was recorded during summer (from July 19th to August 31st of 2015), only 10% of the instances correspond to clear sky conditions. The input features and the output were normalized using min-max normalization. Out of the total dataset, 70% was used for training, while the rest was equally split into validation and test sets. Further information regarding the feature selection, classification, and preparation of the input can be found in [3].

2.4.2 PV Energy Yield Model

Due to confidentiality reasons, a nonlinear auto-regressive network with exogenous inputs (NARX) was used as the PV Eyield model for the case study demonstrations, rather than the physics-based model of [19]. The NARX network was chosen mainly

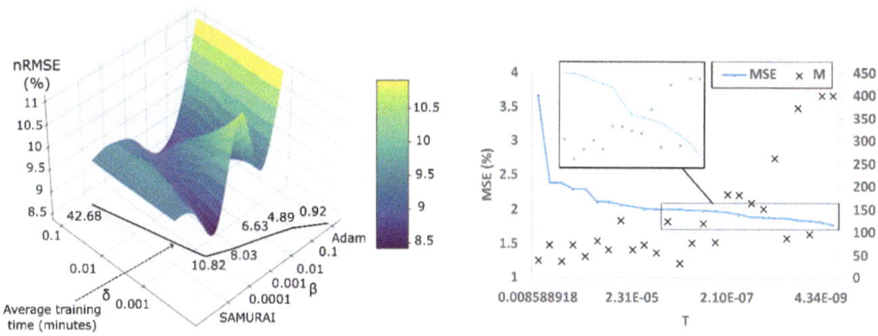

(a) (b)

Fig. 2.9 (**a**) Sensitivity analysis of meta-parameters β and δ for the optimization of the NARX network corresponding to CC 2. (**b**) SAMURAI annealing schedule during training of the NARX network corresponding to CC 3

due to its auto-regressive output, which fits the cumulative nature of Eyield. Further details about the structure and advantages of the NARX network can be found in [3].

A NARX network is separately trained for each cloud class. The parameters to be optimized are the weight synapses of the NARX networks. Therefore, the search process is formed as a continuous optimization problem. Minimization of the mean squared error (MSE) is chosen as the objective function. The search space navigation and move range limitation of SAMURAI are determined based on the method suggested in [34], which is a continuous Monte Carlo method that achieves maximum information gain while maintaining movement anisotropy by adaptively narrowing the search as the temperature decreases.

The optimal values of the meta-parameters β and δ vary depending on the CC and the model type. Therefore, it is essential to fine-tune β and δ independently for each case, to fully activate the potential of the proposed hybrid optimization algorithm. Figure 2.9a presents the sensitivity analysis results for the optimization of the NARX network corresponding to CC 2 (Cirrus/Cirrostratus). The root-mean-squared error (RMSE) averaged over the forecasting horizon is used for evaluation. Furthermore, Fig. 2.9a includes the average training times corresponding to some β values, as well as the results of non-hybrid SAMURAI and Adam. Adam has the lowest training time and accuracy due to premature convergence. On the other hand, the training time explodes when Adam is decoupled from SAMURAI, with a simultaneous decrease in the forecasting accuracy. This is mainly caused by the poor local convergence capability of SAMURAI. For most CCs, the optimal performance is recorded for $\beta = 0.001$, which gives enough time to SAMURAI to traverse through all hierarchical levels before switching to Adam. The only exceptions are CCs 6 and 7 (overcast conditions), where β is significantly larger. In these cases, the objective function is simpler, and less temperature steps are needed to find the neighborhood of the global optimum. The optimal value of δ depends on the difficulty of each case.

Figure 2.9b illustrates the annealing schedule of SAMURAI during training of the NARX network corresponding to CC 3 (Altocumulus/Cirrocumulus). The horizontal axis corresponds to the temperature, the vertical axis to the left corresponds to the objective function, and the vertical axis to the right corresponds to the number of inner-loop evaluations. As seen in Fig. 2.9b, the inner-loop evaluations are momentarily increased in critical regions with steeper decreases of the objective function. SAMURAI spends relatively more time at the final temperature steps, indicating that the search is more localized and fine-grained.

2.4.3 Results

The proposed minute-scale short-term PV Eyield forecasting model, presented in Sect. 2.4.2, is compared with four benchmark models: (a) the persistence model, (b) a random forest (RF) model optimized by Ant-Lion Optimization (ALO), proposed in [23] for short-term forecasting of PV generation, (c) a simple Long Short-Term Memory (LSTM) network, and (d) the NARX-based model presented in Sect. 2.4.2, but instead of the hybrid optimization schedule, the model is trained with classic gradient descent optimization. The normalized RMSE is used for evaluation, averaged for all steps of the forecasting horizon:

$$\overline{nRMSE} = \frac{1}{Eyield_{max}} \sqrt{\frac{1}{sN} \sum_{i=1}^{N} \sum_{j=1}^{s} \left(y_{ij} - \hat{y}_{ij}\right)^2} \tag{2.4}$$

where $Eyield_{max}$ is the maximum Eyield the PV module can generate during the forecasting horizon, N is the number of input instances, s is the number of forecasting steps, and y_{ij}, $\hat{y}_{ij}$ are the real and predicted Eyield values, respectively. All models were developed in Python, and experiments were conducted on an Intel(R) Core (TM) i7-8700 CPU (3.20 GHz, 6 cores) desktop computer with 8 GB of RAM.

Table 2.1 presents the average prediction errors of the compared models per CC as well as their aggregated forecasting performance and computational efficiency. The proposed model has the best overall forecasting performance, exhibiting the lowest average errors for four out seven CCs and the most consistent performance across all sky conditions. Compared to the persistence model, the proposed model improves the forecasting accuracy by more than 10% for all cloudy conditions, except for CC 6, which corresponds to completely overcast conditions and includes only 3% of the total input instances. Compared to gradient descent optimization, the proposed hybrid optimization algorithm reduces the forecasting error by an average of 11.05%. However, the improved forecasting accuracy of the proposed model comes at the expense of the total training time. Compared to gradient descent optimization, the proposed hybrid optimization algorithm takes approximately three

Table 2.1 Final forecasting results and computational efficiency of the compared models

Model	CC 1	CC 2	CC 3	CC 4	CC 5	CC 6	CC 7	Avg	Training time (min)	Inference time (ms)
	$\overline{nRMSE}$ (%)									
Persistence	8.20	11.82	10.07	11.47	8.83	4.37	5.35	9.01	–	–
RF-ALO	8.21	9.03	9.90	8.41	8.08	4.70	5.09	8.01	61.52	4.92
LSTM	7.66	9.86	8.64	8.53	7.12	6.26	5.08	7.74	64.47	37.21
NARX	9.40	11.07	8.88	9.43	7.49	4.53	4.98	8.42	35.53	39.20
NARX-hybrid (proposed)	7.56	8.55	8.13	9.09	7.49	4.41	4.71	7.49	105.4	39.20

times longer to train, to avoid premature convergence and achieve an optimal final set of parameters. Regarding the inference time, the proposed model needs less than 40ms to generate a set of 900 PV Eyield forecasts, corresponding to the 15-minute forecasting horizon with a 1-second resolution. This leaves enough time for GSI acquisition and processing, even at second-scale forecasting intervals, especially if considering that parts of the proposed hybrid optimization algorithm can easily be translated to less computationally requiring programming languages, such as C.

2.5 Mid-Term PV Power Forecasting

Mid-term PV power forecasting covers intraday predictions, the forecasting horizon of which ranges between 1 hour and several hours ahead. Intraday forecasts are of particular importance for modern power systems, as they bridge day-ahead planning operations with real-time microgrid control. An intermediate energy management level can be introduced with mid-term forecasting, which contributes to mitigating the negative effects of low-quality day-ahead forecasts while offering proactive control that reduces the challenges and risks of real-time decision-making. Mid-term PV power forecasting is commonly used for energy management tasks such as demand response and battery scheduling. Furthermore, mid-term PV power forecasting is essential for modern electricity markets and intraday energy trading.

Minute-scale forecasts are still advantageous in mid-term forecasting horizons; however, the forecasting resolution rarely needs to exceed 1 minute. Thus, the temporal smoothing effect becomes more pronounced in mid-term forecasting due to the coarser granularity of the generated forecasts. In this respect, mid-term forecasting is less challenging compared to short-term forecasting, and optimal forecasting performance can be achieved by data-driven models.

Similarly to forecasting PV generation in the short term, remote sensing image data constitutes the optimal input option for mid-term PV power forecasting. Satellite images are preferred over GSIs, as their temporal resolution aligns more effectively with forecasting horizons that exceed 1 hour [7]. Up to 6 hours ahead, satellite images provide good insight into future sky conditions for areas spanning several dozen kilometers. In the absence of satellite images, on-site PV-related measurements and Numerical Weather Predictions (NWPs) are the most common input alternatives. While on-site PV-related measurements provide sufficient information for predictions stretching up to 2 hours ahead, they fail to provide reliable insight further into the future. Therefore, NWPs should be preferred for longer forecasting horizons [5]. Nevertheless, combining satellite images with on-site PV-related measurements and NWPs into hybrid input sets can be the most advantageous approach for data-driven mid-term PV power forecasting models. In this section, however, we focus on mid-term minute-scale PV power forecasting models that are solely based on on-site PV-related measurements.

2.5.1 PV Power Time-Series Decomposition

The performance of data-driven mid-term PV power forecasting models highly depends on the quality of the input data used. When satellite images and NWPs are not available, on-site PV-related measurements remain as the only viable input option. However, as on-site PV-related measurements do not provide any insight on future weather conditions, it is difficult for data-driven mid-term PV power forecasting models to accurately predict the future trend and fluctuations of the PV power time series. In this regard, time-series analysis and preprocessing techniques are of particular importance. Signal decomposition is often used to decompose the input time series into sub-components that carry information about different properties of the original signal. This way, it is easier for the data-driven model to identify these properties and predict the future behavior of the target time series. PV power generation exhibits strong temporal interdependencies at various frequency bands, such as seasonal trends, diurnal cycles, and fast-dynamic fluctuations in the presence of cloud formations. Therefore, wavelet transformation methods are commonly used for decomposing the PV power time series into sub-series that contain information related to different frequency bands.

2.5.2 Weather Classification

Input classification for mid-term PV power forecasting is equally important as when forecasting PV generation in the short term. However, owing to the longer forecasting horizon and the lower temporal resolution of mid-term PV power forecasting, a coarse-grained classification based on the prevailing weather conditions is adequate. In such a coarse-grained classification approach, days are usually divided into sunny, cloudy, and overcast days. As mid-term PV power forecasting models are separately calibrated for each weather class, the optimal choice for the total number of weather classes becomes a trade-off between forecasting accuracy and computational efficiency.

2.5.3 Model Structure and Optimization

Hybrid models have exhibited the most promising results for mid-term PV power forecasting, as the limitations of each sub-component of the hybrid model can be mitigated to some extent by the advantages of the rest of the hybrid model's sub-components. Hybrid forecasting models are further enhanced when combined with techniques such as signal preprocessing, feature engineering, and metaheuristics-based fine-tuning. Especially when combined with signal decomposition, hybrid models achieve optimal performance when they are nonhomogeneous, i.e., when a different sub-model is employed for each of the sub-series of the decomposed original signal. Each of the generated sub-series often exhibits different characteristics; thus, a sub-model should be chosen carefully to optimally fit the specific characteristics of the corresponding sub-series.

The main drawback of nonhomogeneous hybrid models is their excessive computational complexity [7]. While this may not be an issue for big PV power plant owners, it can be an important limitation for owners of local small-scale rooftop PV systems. Furthermore, the optimization of nonhomogeneous hybrid PV power forecasting models is also a challenge due to the excessive number of parameters to be optimized. Component-wise or layer-wise fine-tuning, i.e., individual optimization of each sub-component, leads to suboptimal configurations. On the other hand, optimizing each sub-component of a hybrid model with respect to the optimization of the rest of the sub-components can lead to significant improvements in the forecasting accuracy. We refer to this process as "co-optimization." The excessive number of parameters to be co-optimized in nonhomogeneous hybrid models; however, makes exhaustive search and grid search methods unfeasible options for co-optimization. Therefore, we advocate the use of sophisticated metaheuristic algorithms for the co-optimization of mid-term hybrid PV power forecasting models, to limit the additional computational cost of the overall forecasting approach while maintaining scalability.

2.5.4 Hours-Ahead Probabilistic Extension

Extending the forecasting horizon up to several hours ahead inevitably increases the forecasting uncertainty, especially in the absence of satellite images and NWPs. Therefore, probabilistic forecasting models are frequently preferred for mid-term PV power forecasting, as they offer a wider perspective of the predictive outcome by quantifying the forecasting uncertainty [8]. The optimal type of probabilistic output depends on the target application, the optimization approach, and the forecasting horizon. For example, for day-ahead scheduling of energy systems with stochastic optimization, PV power forecasts could be issued in the form of scenarios [29]. However, when forecasting PV generation up to several hours ahead, we advocate the usage of prediction intervals (PIs), which offer valuable insight on the upper and lower bounds of the target variable while maintaining high levels of interpretability and computational efficiency.

2.5.5 Mid-Term PV Power Forecasting Implementation

In this section, we present the implementation details and results generated by two data-driven mid-term PV power forecasting models which are solely based on on-site PV-related measurements. The first model, introduced in [30], is a data-driven hybrid non-homogeneous structure based on signal decomposition and co-optimization, developed to generate 1 hour ahead forecasts. The second model, introduced in [8], is a data-driven hybrid probabilistic PV power forecasting model, suitable for extending the forecasting horizon up to several hours ahead. Henceforth, we will refer to the first model as the deterministic mid-term PV power forecasting model, and to the second model as the probabilistic mid-term PV power forecasting model.

A schematic overview of the deterministic mid-term PV power forecasting model is presented in Fig. 2.10a [30]. The PV power generation time series is decomposed into an approximation sub-series and three residual sub-series, using wavelet packet decomposition (WPD) and single-branch reconstruction. Distinct data-driven sub-models are employed for the approximation and residual sub-components, to allow for more effective analysis and modeling. The seasonal long-term temporal autocorrelations and strong intraday patterns of PV power generation constitute LSTM an optimal data-driven architecture for handling the approximation sub-component. A straightforward choice would be to use LSTM networks for the residuals as well. However, besides not being the most computationally efficient option, this strategy can also lead to sub-optimal forecasting performance. The residuals of the PV power time series exhibit increased stochasticity with negligible long-term temporal autocorrelations. Therefore, RFs are employed for handling the residual components of the PV power time series, to reduce the required inference time and computational requirements. The outputs of each sub-model are linearly combined to provide the final PV power forecast. The Ant-Lion Optimization (ALO) algorithm is employed for the co-optimization of the parameters of the deterministic mid-term PV power forecasting model. Further information about the deterministic mid-term PV power forecasting model can be found in [30].

The deterministic mid-term PV power forecasting model is evaluated on real measurements extracted from a 1.2 MW PV power plant, located in Greece. Measurements were conducted at a 1-minute resolution from the 20th of July 2022 until the 16th of July 2023. The temporal resolution of the forecasts is kept as high as possible, i.e., at 1 minute, to reduce the temporal smoothing effect and evaluate the proposed mid-term PV power forecasting model at the maximum level of difficulty. Furthermore, an ablation study is conducted to evaluate the importance of each component of the proposed hybrid mid-term PV power forecasting model. Initially, WPD is removed, and the original PV power generation time series is directly fed to two non-hybrid models: a simple RF and an LSTM. Then, WPD is applied to decompose the original PV power time series into four sub-series, as shown in Fig. 2.10a. Two homogeneous hybrid approaches are tested, employing solely RFs (WPD-RF) or LSTM networks (WPD-LSTM) for all sub-components. Each sub-model undergoes component-wise optimization using grid search. Finally, the deterministic mid-term PV power forecasting model is evaluated with component-wise optimization (WPD-LSTM-RF) and ALO-based co-optimization (WPD-LSTM-RF-ALO). Further implementation details can be found in [30].

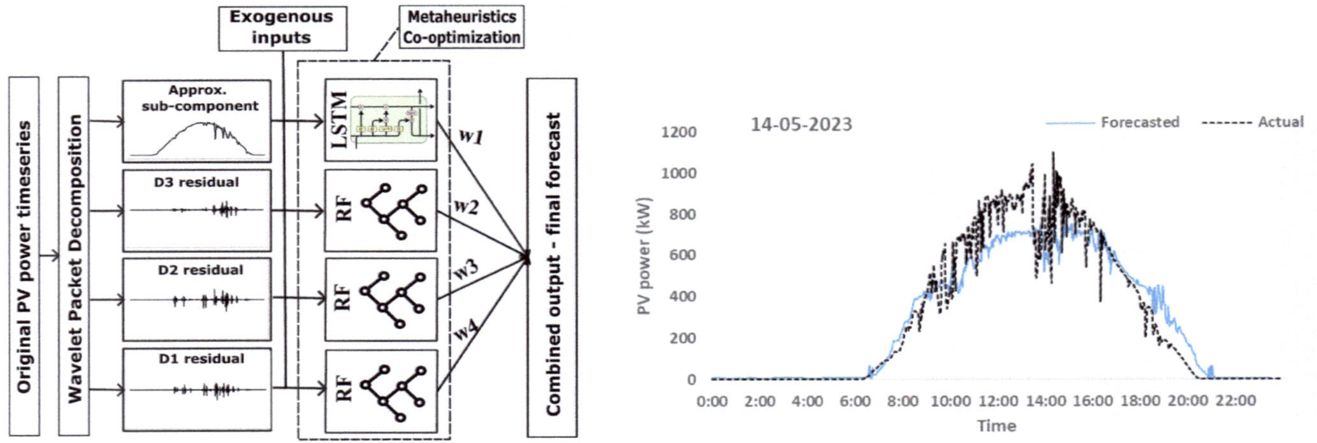

(a) (b)

Fig. 2.10 (**a**) Proposed deterministic PV power forecasting model [30] (**b**) PV power generation, as predicted by the deterministic PV power forecasting model, during a cloudy day

Table 2.2 Forecasting performance and computational efficiency of the compared deterministic mid-term PV power forecasting models, under different weather conditions

	Sunny	Cloudy	Overcast		
Model	$nRMSE$ (%)			Training time (s)	Inference time (μs)
LSTM	9.97	13.44	16.74	40.37	75.42
RF	8.64	13.32	19.37	62.11	7.53
WPD-LSTM	9.41	13.16	15.76	67.59	259.95
WPD-RF	8.12	13.05	15.88	126.54	15.68
WPD-LSTM-RF	8.67	13.08	15.95	116.78	66.13
WPD-LSTM-RF-ALO	8.07	12.83	14.86	102.87	69.47

Table 2.2 presents the final forecasting performance under different weather conditions as well as the average computational efficiency of all compared models. The non-hybrid LSTM and RF models generally underperform under the more challenging cloudy and overcast conditions. The addition of WPD and the use of hybrid approaches (WPD-LSTM, WPD-RF, WPD-LSTM-RF) significantly reduce the forecasting error, with nRMSE improvements ranging from 2% to 18%. Fully co-optimizing the non-homogeneous WPD-LSTM-RF model with ALO reduces the nRMSE by 6.92%, 1.91%, and 6.83% under sunny, cloudy, and overcast conditions, respectively. Compared to the second-best model of each weather condition, the proposed WPD- LSTM-RF-ALO model reduces the nRMSE by 0.62%, 1.69%, and 5.71% under sunny, cloudy, and overcast conditions, respectively. Thus, the more challenging the weather conditions, the bigger the necessity of the proposed co-optimized nonhomogeneous hybrid approach.

The average training time obviously increases with the complexity of the forecasting model. On the other hand, the memory requirements and inference time of RFs are significantly lower compared to those of LSTM networks. Therefore, replacing LSTMs with RFs in handling the residuals not only improves the forecasting performance but also significantly reduces the average inference time. The computational efficiency of the proposed model is further improved due to co-optimization with the ALO algorithm. ALO is completely scalable, in contrast to grid search which quickly explodes when fine-tuning more than six parameters. For a total set of 15 parameters, grid search co-optimization becomes completely unfeasible, as it would require approximately 5 years to explore all combinations generated for just 3 values per parameter.

Figure 2.10b presents an example of the PV power curve predicted by the proposed mid-term PV power forecasting model during a cloudy day. The forecasted PV power curve generally follows the actual trend and manages to capture ramp events to some extent, despite the unfavorable weather conditions and the fact that only numerical on-site measurements are provided as input. The time shift between the predicted and the actual PV power curves ranges from 10 to 38 minutes, which is sufficiently smaller than the forecasting horizon. PV power generation is slightly underestimated during peak hours, which can easily be mitigated by adding temporal and peak-hour features to the input set.

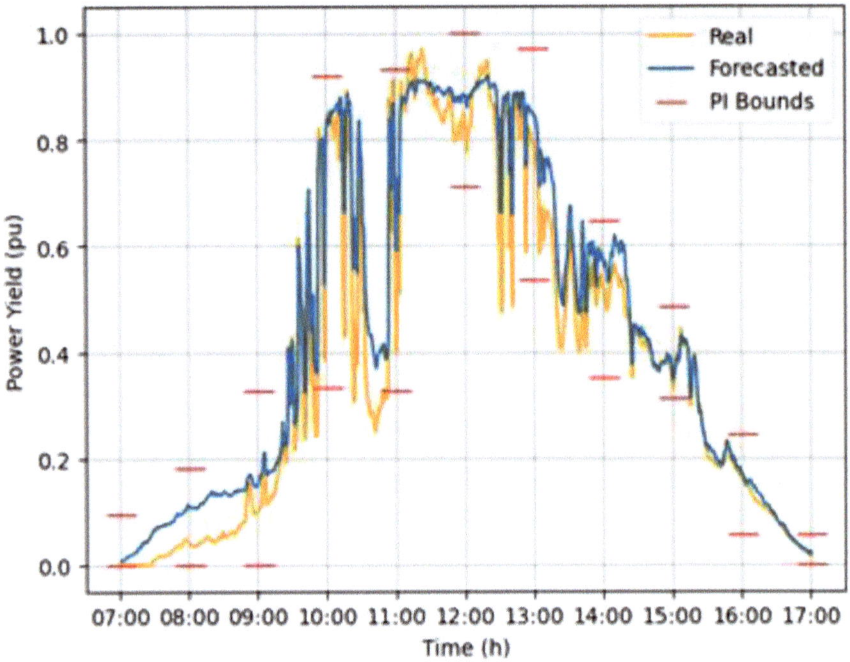

(a)

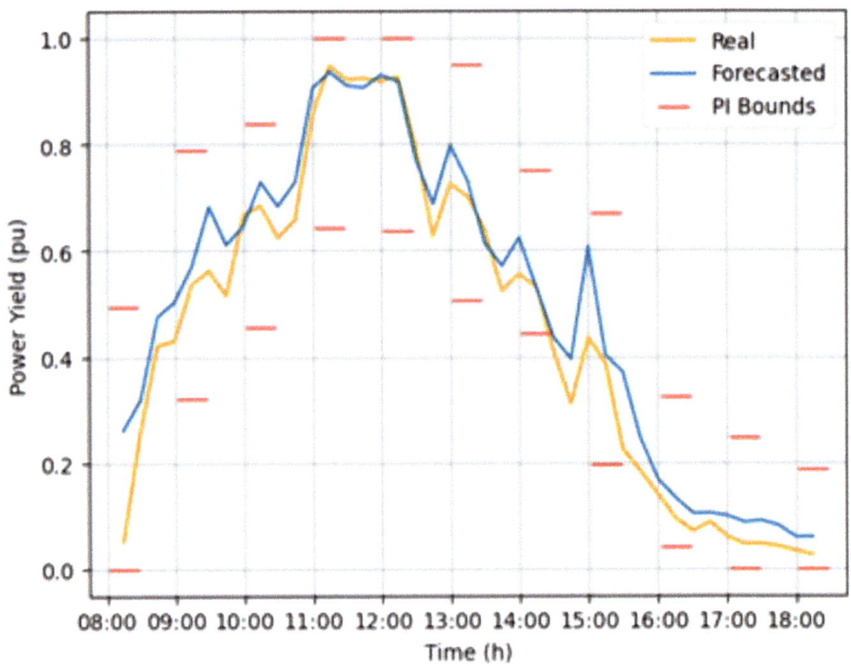

(b)

Fig. 2.11 Hourly PIs and intra-hour point predictions generated by the probabilistic mid-term PV power forecasting model under cloudy conditions. (**a**) PVPP 1. (**b**) PVPP 2

The probabilistic mid-term PV power forecasting model is based on LSTM networks, which handle the original PV power time series to generate initial minute-scale point predictions. PIs are then constructed with the bootstrap technique in an hourly resolution. Bootstrap sampling is applied on daily subsets of historical data, to maintain the diurnal temporal autocorrelation of the PV power time series. The generated PIs are then further optimized by an improved Chicken Swarm Optimization algorithm, which is coupled with a prey-predator (PP) mechanism. The aim is to increase the sharpness of the PIs while maintaining their coverage above the specified confidence level. Further information about the probabilistic mid-term PV power forecasting model can be found in [8].

The probabilistic mid-term PV power forecasting model is evaluated on two distinct case studies. The first case study refers to a 1.5MW PV Power Plant (PVPP 1), in which measurements are extracted at a 1-minute resolution. The second case study refers to a 11.9MW PV Power Plant (PVPP 2), in which measurements are extracted at a 15-minute resolution. For both case studies, the proposed probabilistic mid-term PV power forecasting model achieves superior forecasting performance in comparison with all benchmark models, regardless of the prevailing weather conditions. PIs were generated for 90%, 95%, and 99% confidence levels. In all cases, the proposed model maximizes the sharpness of the generated PIs, while maintaining their average coverage above the specified confidence level. Furthermore, the proposed model generates reliable point predictions that provide valuable insight regarding the intra-hour PV power fluctuations. Figure 2.11 illustrates an example of the hourly PIs and the intra-hour point predictions generated by the probabilistic mid-term PV power forecasting model for both case studies under cloudy conditions. Further implementation details and results regarding the probabilistic mid-term PV power forecasting model can be found in [8].

2.6 Conclusions

In this chapter, we summarized the basic principles and challenges of forecasting PV generation in minute-scale resolutions. Especially for short-term forecasting horizons, we believe that PV Eyield forecasting models will become extremely relevant in the coming years, due to the continuous integration of small-scale building PV systems and the consequent increasing need for smart energy management frameworks. As the forecasting challenge increases with the temporal resolution of the forecasts, we advocate the use of sophisticated physics-based PV models, as long as the forecasting horizon allows for their computational intensity. Regarding the input data, remote sensing information provided by simple camera setups and satellites offer significant insight on future sky conditions. Regardless of the modeling approach, optimization plays a crucial role in the overall forecasting accuracy, computational efficiency, and scalability, as demonstrated both for short-term and mid-term forecasting horizons. Further demonstrations can be found in related work we have published over the last decade, provided in the references.

We believe that the main challenge regarding minute-scale PV Eyield forecasting will be to limit the required computational resources and develop models that are fully adaptive and resilient. PV Eyield models should be accessible and affordable for everyday end-users, to facilitate the smoother integration of PV systems into smart microgrids and energy communities, reduce the average energy costs and CO_2 emissions, and promote a more sustainable lifestyle. Furthermore, in the era of artificial intelligence, green computing solutions have become more crucial than ever. Thus, our aim with this chapter was not only to provide the basis for accurate and efficient minute-scale PV Eyield forecasting but also to demonstrate the importance of the collaborative interaction between different disciplines to achieve optimal forecasting results.

References

1. Aarts, E.E., Laarhoven, V.: Statistical cooling: a general approach to combinatorial optimization problems. Philips J. Res. **40**, 193–226 (1985)
2. Anagnostos, D., Goverde, H., Catthoor, F., Soudris, D., Poortmans, J.: Systematic cross-validation of photovoltaic energy yield models for dynamic environmental conditions. Solar Energy **155**, 698–705 (2017)
3. Anagnostos, D., Schmidt, T., Cavadias, S., Soudris, D., Poortmans, J., Catthoor, F.: A method for detailed, short-term energy yield forecasting of photovoltaic installations. Renew. Energy **130**, 122–129 (2019)
4. Anagnostos, D., Schmidt, T., Goverde, H., Kalisch, J., Catthoor, F., Soudris, D.: Pv energy yield nowcasting combining sky imaging with simulation models. In: 31st European Photovoltaic Solar Energy Conference and Exhibition, pp. 1552–1555 (2015)
5. Bacher, P., Madsen, H., Nielsen, H.A.: Online short-term solar power forecasting. Solar Energy **83**(10), 1772–1783 (2009)
6. Barbieri, F., Rajakaruna, S., Ghosh, A.: Very short-term photovoltaic power forecasting with cloud modeling: a review. Renew. Sustain. Energy Rev. **75**, 242–263 (2017)
7. Bazionis, I.K., Kousounadis-Knousen, M.A., Georgilakis, P.S., Shirazi, E., Soudris, D., Catthoor, F.: A taxonomy of short-term solar power forecasting: Classifications focused on climatic conditions and input data. IET Renewable Power Generation **17**(9), 2411–2432 (2023)
8. Bazionis, I.K., Kousounadis-Knousen, M.A., Katsigiannis, V.E., Catthoor, F., Georgilakis, P.S.: An advanced hybrid boot-lstm-icso-pp approach for day-ahead probabilistic pv power yield forecasting and intra-hour power fluctuation estimation. IEEE Access **12**, 43704–43720 (2024)
9. Catthoor, F., de Man, H., Vandewalle, J.: Samurai: A general and efficient simulated-annealing schedule with fully adaptive annealing parameters. Integration **6**(2), 147–178 (1988)
10. Chang, X., Li, W., Zomaya, A.Y.: A lightweight short-term photovoltaic power prediction for edge computing. IEEE Transactions on Green Communications and Networking **4**(4), 946–955 (2020)
11. Chen, X., Du, Y., Lim, E., Fang, L., Yan, K.: Towards the applicability of solar nowcasting: A practice on predictive pv power ramp-rate control. Renewable Energy **195**, 147–166 (2022)
12. Chowdhury, M.G., Goossens, D., Goverde, H., Catthoor, F.: Experimentally validated cfd simulations predicting wind effects on photovoltaic modules mounted on inclined surfaces. Sustainable Energy Technologies and Assessments **30**, 201–208 (2018)

13. Chowdhury, M.G., Goverde, H., Manganiello, P., Voroshazi, E., Poortmans, J., Catthoor, F.: Sensitivity analysis of the effect of forced convection on photovoltaic module temperature and energy yield. In: 2019 IEEE 46th Photovoltaic Specialists Conference (PVSC), pp. 1237–1241 (2019)
14. Diagne, M., David, M., Lauret, P., Boland, J., Schmutz, N.: Review of solar irradiance forecasting methods and a proposition for small-scale insular grids. Renew. Sustain. Energy Rev. **27**, 65–76 (2013)
15. Dissawa, L.H., Godaliyadda, R.I., Ekanayake, P.B., Agalgaonkar, A.P., Robinson, D., Ekanayake, J.B., Perera, S.: Sky image-based localized, short-term solar irradiance forecasting for multiple PV sites via cloud motion tracking. Int. J. Photoenergy **2021**(1), 9973010 (2021)
16. Goossens, D., Goverde, H., Catthoor, F.: Effect of wind on temperature patterns, electrical characteristics, and performance of building-integrated and building-applied inclined photovoltaic modules. Solar Energy **170**, 64–75 (2018)
17. Goverde, H., Anagnostos, D., Herteleer, B., Govaerts, J., Baert, K., Aldalali, B., Catthoor, F., Driesen, J., Poortmans, J.: Model requirements for accurate short term energy yield predictions during fast-varying weather conditions. In: 31st European Photovoltaic Solar Energy Conference and Exhibition, pp. 1556–1559 (2015)
18. Goverde, H., Goossens, D., Govaerts, J., Catthoor, F., Baert, K., Poortmans, J., Driesen, J.: Spatial and temporal analysis of wind effects on PV modules: consequences for electrical power evaluation. Solar Energy **147**, 292–299 (2017)
19. Goverde, H., Govaerts, J., Baert, J., Catthoor, F., Driesen, J., Poortmans, J.: Optical-thermal-electrical model for a single cell PV module in non-steady-state and non-uniform conditions build in spice. In: 28th European Photovoltaic Solar Energy Conference and Exhibition, pp. 3291–3295 (2013)
20. Goverde, H., Herteleer, B., Anagnostos, D., Köse, G., Goossens, D., Aldalali, B., Govaerts, J., Baert, K., Catthoor, F., Driesen, J., Poortmans, J.: Energy yield prediction model for PV modules including spatial and temporal effects. In: 29th European Photovoltaic Solar Energy Conference and Exhibition, pp. 3292–3296 (2014)
21. Hasenbalg, M., Kuhn, P., Wilbert, S., Nouri, B., Kazantzidis, A.: Benchmarking of six cloud segmentation algorithms for ground-based all-sky imagers. Solar Energy **201**, 596–614 (2020)
22. Heinle, A., Macke, A., Srivastav, A.: Automatic cloud classification of whole sky images. Atmos. Meas. Tech. **3**(3), 557–567 (2010)
23. Ibrahim, I.A., Hossain, M.J., Duck, B.C.: An optimized offline random forests-based model for ultra-short-term prediction of PV characteristics. IEEE Trans. Ind. Inform. **16**(1), 202–214 (2020)
24. Kazantzidis, A., Tzoumanikas, P., Bais, A., Fotopoulos, S., Economou, G.: Cloud detection and classification with the use of whole-sky ground-based images. Atmos. Res. **113**, 80–88 (2012)
25. Kazantzidis, A., Tzoumanikas, P., Blanc, P., Massip, P., Wilbert, S., Ramirez-Santigosa, L.: Short-term forecasting based on all-sky cameras. In: Kariniotakis, G. (ed.) Renewable Energy Forecasting, Woodhead Publishing Series in Energy, pp. 153–178. Woodhead Publishing (2017)
26. Kirkpatrick, S., Gelatt, C.D., Vecchi, M.P.: Optimization by simulated annealing. Science **220**(4598), 671–680 (1983)
27. Kleissl, J.: Solar Energy Forecasting and Resource Assessment. Elsevier, Amsterdam (2013)
28. Kousounadis-Knousen, M.A., Bakovasilis, A., Catthoor, F., Georgilakis, P.S.: Resource-efficient PV energy yield nowcasting with sky images: a hybrid global annealing schedule. In: 41st European Photovoltaic Solar Energy Conference and Exhibition (2024)
29. Kousounadis-Knousen, M.A., Bazionis, I.K., Georgilaki, A.P., Catthoor, F., Georgilakis, P.S.: A review of solar power scenario generation methods with focus on weather classifications, temporal horizons, and deep generative models. Energies **16**(15) (2023)
30. Kousounadis-Knousen, M.A., Bazionis, I.K., Georgilaki, A.P., Georgilakis, P.S.: A fully co-optimized hybrid approach for short-term pv power forecasting that addresses the challenges of diurnal trends and residuals. In: 14th Mediterranean Conference on Power Generation Transmission, Distribution and Energy Conversion (2024)

31. Lin, F., Zhang, Y., Wang, J.: Recent advances in intra-hour solar forecasting: A review of ground-based sky image methods. Int. J. Forecasting **39**(1), 244–265 (2023)
32. Schmidt, T.: High resolution solar irradiance forecasts based on sky images. Ph.D. Thesis, Fakultät für Mathematik und Naturwissenschaften, Carl von Ossietzky Universität Oldenburg (2017)
33. Tsourounis, D., Kastaniotis, D., Theoharatos, C., Kazantzidis, A., Economou, G.: Sift-CNN: when convolutional neural networks meet dense sift descriptors for image and sequence classification. J. Imaging **8**(10), 256 (2022)
34. Vanderbilt, D., Louie, S.G.: A monte carlo simulated annealing approach to optimization over continuous variables. J. Comput. Phys. **56**(2), 259–271 (1984)
35. Watari, D., Taniguchi, I., Goverde, H., Manganiello, P., Shirazi, E., Catthoor, F., Onoye, T.: Multi-time scale energy management framework for smart pv systems mixing fast and slow dynamics. Appl. Energy **289**, 116671 (2021)
36. West, S.R., Rowe, D., Sayeef, S., Berry, A.: Short-term irradiance forecasting using skycams: motivation and development. Solar Energy **110**, 188–207 (2014)

Chapter 3
Accurate Energy Yield Forecasting for Wind Turbine Parks

Dimitris Michos, Andreas Kazantzidis, Markos A. Kousounadis-Knousen, Ioannis K. Bazionis, Pavlos S. Georgilakis, Michael Daenen, and Francky Catthoor

Just as for PV installations (see Chap. 2), the past decades have shown a rapid yearly growth in wind turbine plants. So together with PV, it is the main driver for the society-level transition to renewable technologies. In 2023, the global cumulative wind turbine (WT) capacity increased by 117 GW, reaching a total of about 1 TW. Continuous advancements in WT technologies have enhanced their cost-benefit ratio by reducing manufacturing costs while increasing Eyield efficiency.

Similar again to the PV installations, WT power generation is characterized by high variability and volatility, mainly due to the strongly varying wind fields both due to thermal fluctuations and terrain effects. Especially in forecasting horizons where the averaging out effect is less applicable, the intermittency and volatility of wind power can lead to forecasts of significant uncertainty. With wind power penetrating the energy systems more and more every year over the last decade, researchers have focused on quantifying this forecasting uncertainty. This variability can have severe effects on the effective operation of modern wind parks. Therefore, broader integration of such systems again necessitates the introduction of smart energy management systems (EMSs) (see Chap. 5), which in turn rely on accurate forecasts of Eyield generation.

D. Michos · A. Kazantzidis (✉)
Laboratory of Atmospheric Physics, Physics Department, University of Patras, Patras, Greece
e-mail: akaza@upatras.gr

M. A. Kousounadis-Knousen · I. K. Bazionis · P. S. Georgilakis
School of Electrical and Computer Engineering, National Technical University of Athens, Athens, Greece

M. Daenen
IMOMEC, Hasselt, Belgium

F. Catthoor
Zografou Campus, National Technical University of Athens, Athens, Greece

F. Catthoor et al. (eds.), *Energy Production, Load and Battery Management Framework with Supporting Methods for Smart Microgrids*,
https://doi.org/10.1007/978-3-031-92025-7_3

However, optimal EMSs include multiple timescales (see Chap. 5 also); thus, forecasts of WT Eyield generation need to be issued for different forecasting horizons, similar to the PV case in Fig. 2.1. Among these, minute-scale forecasts enable specific real-time EMS operations, such as ramp rate control, demand balancing, and battery scheduling [1, 3]. In this chapter, we will address how to better deal with this variation both for ultra-short-term (10–30 minutes ahead) and short-term (a few hours ahead) forecasting timescales.

3.1 Physics-Based Ultra-Short-Term Wind Energy Forecasting

Wind energy forecasting is crucial for optimizing energy management by improving the use and storage of diverse energy sources and enabling rapid responses to extreme conditions [30]. In energy sectors characterized by dynamic demand and storage capabilities, predicting energy production over short time horizons is essential for the effective implementation of dynamic pricing systems based on smart grids, similar to those used for longer time frames [24, 27, 29]. This growing demand for ultra-short-term wind energy forecasting is driven by the need to minimize power losses, stabilize electricity prices, and achieve cost savings through improved energy storage and management. In the context of this book, it is also crucial to drive the production input for the load-store balancing and the holistic energy management of the micro-grid (see Chap. 5).

3.1.1 Problem Formulation and Context

Accurate and reliable wind energy forecasting plays a pivotal role in integrating wind power into the grid and ensuring stability. This is especially critical in ultra-short-term forecasts, which focus on predictions within a 30-minute window, as they help to optimize energy management strategies by leveraging the benefits of precise forecasts. This is even more needed to deal with real-world wind farm (WF) modelling limitations which have emerged. These derive from economic factors related to equipment update and the new methods which have been developed to support modern WT types. Predicting the Energy Yield (E-yield) of a WT is particularly challenging due to the chaotic nature of wind movement over short timescales and in complex terrains. Wind forecasting models are generally categorized into statistical models and physics-based models. Statistical data-driven models, including Artificial Neural Network (ANN), are valued for their fast and accurate predictions suitable for operational use [12, 36]. These models can directly predict the E-yield or wind speed, as seen in [43]. However, they require extensive datasets for training and must be tailored to specific locations, making them much less suited to changing conditions. Established wind farms (WFs) with minimal infrastructural changes can provide the necessary data, but new WFs need to start measurement campaigns well before the installation of the first turbine. Furthermore, these models need to be retrained when site or environmental changes occur, such as new turbines or neighbouring structures (within km range), which will occur quite often in an urban environment. This leads to costly and time-consuming upgrades which can also disrupt proper operations.

In contrast, physics-based models, like computational fluid dynamics (CFD) models like the ones presented by [32, 37, 41, 42] and [34], offer detailed estimations of wind fields and are less affected by environmental changes compared to statistical models. Wind flows are highly sensitive to terrain features and obstacles, particularly near buildings and urban areas [39, 46]. Even minor terrain modifications, such as constructing a small shed near a WT, can significantly impact the wind field [25]. The complex interactions between three-dimensional wind flows and turbine blades mean that E-yield can be sensitive to terrain changes, especially with elevation fluctuations. However, the high computational demands of physics-based CFD models, which often involve long computation times for time-dependent simulations, limit their application for real-time online ultra-short-term forecasting where rapid predictions are needed.

Our research in the past 5 years has led to an innovative physics-based model designed for ultra-short-term WT power forecasting. The presented model can be separated in two parts, the first which is a wind spatial extrapolation and forecasting model that leverages CFD principles with stationary equations to address the unique challenges posed by complex terrains (see Sect. 3.1), and the second part which is the power forecasting model that takes advantage of a novel, yet simple method to detect the power production latency of a given WT, accounting for persistence in wind conditions (see Sect. 3.2). By spatially extrapolating limited wind LIDAR measurements, the CFD model named WiSpEx, is able to generate high-resolution input datasets that enhance the accuracy of ultra-short-term forecasts [2], even in complex environments. Recognizing the impact of complex terrain on LIDAR measurement accuracy [7, 18, 19], our research has also examined the model's forecasting performance under various conditions through simulations conducted over the CRES site in Greece near to Athens.

The final part of our modeling approach involves WEEL that handles the conversion of simulated wind speeds into power through a continuous power curve approximation, introducing a novel method for detecting power extraction latency in WTs. This approach can empower small, low-cost WFs to optimize their existing setups while enabling further research through the use of physics-based synthetic datasets.

By prioritizing the physics of wind behavior, our novel approach reduces dependency on terrain-specific adjustments, allowing the model to retain the accuracy of traditional physics-based forecasts at a significantly reduced computational cost. It extrapolates the wind field from the measurement locations at the WT over the whole WF (or area of interest). This enables fast and reliable predictions, promoting broader operational use of CFD in wind forecasting and encouraging the development of hybrid models that combine the strengths of statistical and physics-based approaches. Such hybrid models could potentially overcome the limitations of each individual method, delivering more accurate and reliable ultra-short-term wind energy forecasts and fostering further innovation in the field.

In our main experimental study, an attempt to test the limits of CFD simulations to recreate a given wind field over a complex terrain from scarce data, in a sufficiently fast and accurate way, was made. In many occasions, a WF may not be able to withstand the cost of buying and/or maintaining the equipment needed to collect large amounts of quality data, like in our case study. The CRES site, situated in the complex terrain of Lavrio, Greece, presents a unique challenge for wind resource assessment. Wind LIDAR measurements were provided by a specific wind LIDAR campaign conducted at CRES between October 22nd and 24th, 2010, with temporary use of the equipment. Wind measurements were obtained at nine different heights (40 m, 54 m, 78 m, 100 m, 120 m, 140 m, 160 m, 180 m, and 200 m). The potential underestimation of wind speeds and turbulence intensity by the LIDAR is affected by vertical wind shear and atmospheric stability. These limitations underscore the importance of data validation and uncertainty quantification in CFD modeling. It is important to note that the LIDAR wind speed measurements at the CRES site have errors ranging from 4% to 6%, as stated by Bingöl et al. [7]. These errors, likely arising from factors such as terrain slope, vegetation, and atmospheric stability, need to be considered when interpreting the simulation results and assessing their accuracy.

The wind LIDAR measurements, while valuable, lack data close to the ground level. This gap in data, coupled with the inherent errors in LIDAR measurements, underscores the challenges in accurately specifying boundary conditions for CFD simulations. The quality of these boundary conditions directly impacts the reliability and accuracy of simulation outcomes [25].

To address these challenges, our CFD benchmark version is designed to spatially extrapolate wind fields in under 3 minutes, even with limited wind and temperature data. The benchmark model achieves Mean Absolute Percentage Error (MAPE) below 10% for the estimations of wind speed at hub height of the WTs installed in CRES, making it a valuable tool for generating high-resolution wind datasets when critical information is missing. Figure 3.1 illustrates the placement of the LIDAR, WTs, and the surrounding terrain.

The 3D representation of the CRES site and its vicinity, described by a domain volume of approximately 2 km × 2.5 km × 0.6 km is created within the multi-physics framework COMSOL. We have created a subdomain enclosed by the initial, encompassing the locations of the WTs and the LIDAR, which is positioned at the inlet boundary plane of WiSpEx simulations conducted for the inner area. The terrain within the domain is complex, with an average elevation of 66 m and a standard deviation of 33 m. The maximum elevation reaches 136 m, while the minimum is near sea level as stated by [25]

In our study the WT's are not inserted in the simulations to decrease the computational time of the simulations. We conducted Simulations to evaluate the effects of different obstacles and simplifications on wind flow. For instance, a

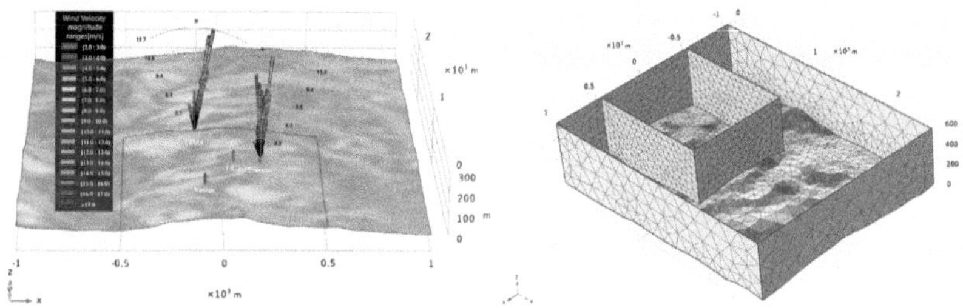

Fig. 3.1 CRES—3d geometry view with LIDAR, WT locations and wind measurement statistics (**left**), meshed geometry (**right**)

non-rotating WT of similar size to those in our case study can produce a wake extending up to 270 m. As shown in [25], we cannot neglect these effects as this would lead to large errors on the E-yield forecasting. For instance, a shed-like structure placed 200 m away from a WT could influence the flow in all directions, potentially extending its effect to the rotor area. These simulations reveal how various obstacles can significantly alter wind flow behavior over large areas.

3.1.2 The Wi.Sp.Ex. Approach

WiSpEx operates under the [25] hypothesis that steadystate solutions of CFD equations can effectively represent the flow over complex, uncontrolled environments under specific conditions. This assumption is rooted in the principles of calculus used in CFD and finite-element theory. The core idea is that the temporal evolution of airflow within a volume (CFD discretized) can be approximated by a series of steady (stationary) states. As the volume increases or the mean velocity of the airflow within the volume rises, a stationary flow should represent the averaged flow over a longer or shorter duration, respectively.

IAS extrapolates a 3D wind field from a 1D field with simplified inlet conditions. EAS extrapolates 1D LIDAR measurements to the 2D inlet wind field of the E-IAS simulation, which aims to capture the effect of terrain characteristics to the flow to. naturally extrapolate Lidar measurements over the wind park. The EAS aims to physically extrapolate the wind profile both vertically and horizontally over the E-IAS inlet plane. To achieve this, a 2D inlet plane wind field was generated, assuming uniform velocity along the X-axis of the inlet and zero vertical velocity component. A ramp function was used to extrapolate LIDAR wind measurements vertically below 40 m, while constant extrapolation was used above 200 m.

Once the EAS results were obtained, a 3D matrix containing the simulated wind velocity vectors at the inlet of the examined area was created. A deviation matrix was then calculated to capture the difference between the EAS simulated LIDAR velocity and the actual LIDAR measurements. Finally, a new matrix was generated to represent the extrapolated inlet wind field, combining the deviation matrix and the extrapolated LIDAR measurements. The vertical velocity component for the E-IAS inlet field was set to zero, similar to the IAS and EAS. The E-IAS simulation was then run over the area of interest,

The results in [25] for the IAS, EAS, and E-IAS (see that paper for more explanations on these 3 different options) compare their performance after being corrected with linear regression, in terms of Mean Absolute Error (MAE), Mean Square Error (MSE), Root Mean Square Error (RMSE) and MAPE. The results are very promising, as shown below. Figures 3.2 and 3.3 present the velocity magnitude plots alongside their respective linear regression scatter plots for the Enercon and NEG WTs. An examination of the velocity magnitudes figures, indicates that wind direction affects the accuracy of the model (Table 3.1).

The E-IAS simulations, producing the final output of the WiSpEx model, demonstrate the ability to transform limited LIDAR measurements into a detailed wind field dataset. As shown in Table 3.1, the MAE for wind velocity estimation is 0.81 m/s for the Enercon WT, 0.76 m/s for the NEG WT, and 0.56 m/s for the Vestas WT.

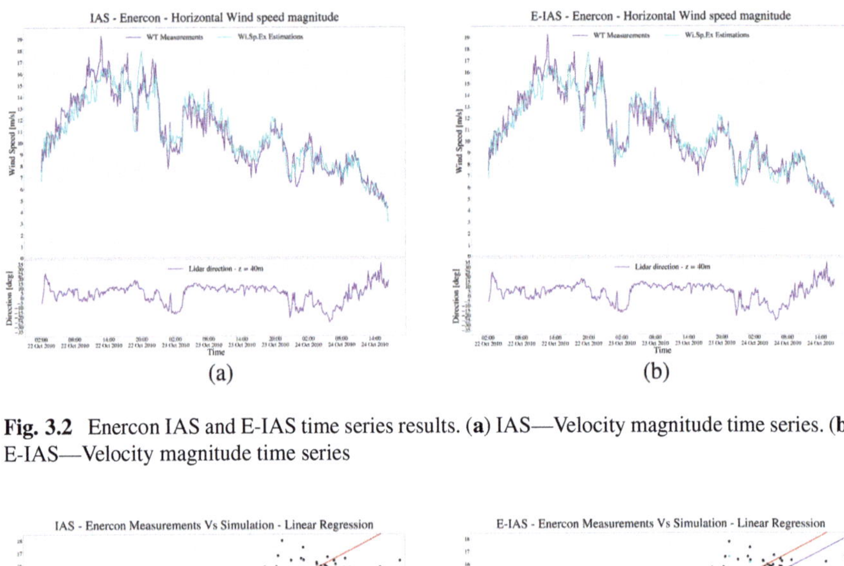

(a) (b)

Fig. 3.2 Enercon IAS and E-IAS time series results. (**a**) IAS—Velocity magnitude time series. (**b**) E-IAS—Velocity magnitude time series

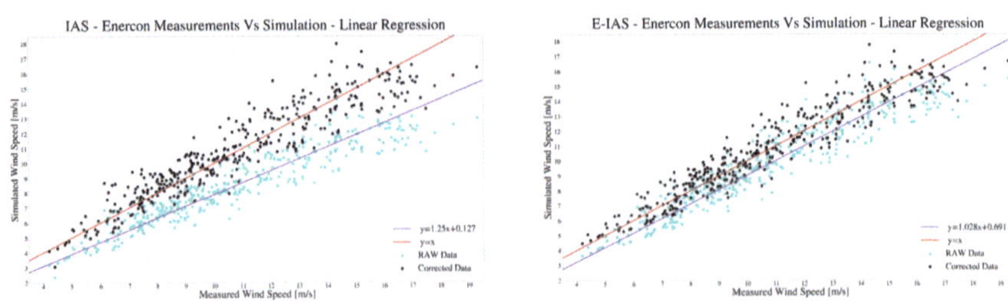

(a) (b)

Fig. 3.3 Enercon IAS and E-IAS scatter plot results. (**a**) IAS—Velocity magnitude linear regression scatter plot. (**b**) E-IAS—Velocity magnitude linear regression scatter plot

Table 3.1 E-IAS – WT Corrected Velocity Magnitude Statistics

	Enercon	NEG
MAE [m/s]	0.81	0.76
MSE [m^2/s^2]	1.08	0.93
RMSE [m/s]	1.04	0.96
MAPE [%]	7.62	7.98

The corresponding MAPE values are 7.62%, 7.98%, and 7.08%, respectively. These results indicate a significant improvement over previous simulations, highlighting the model's efficiency in extrapolating sparse measurements into a comprehensive high-resolution dataset. Specifically, the model successfully reconstructs a 3D wind field at 29,928 locations using only nine vertical measurement points per time step, achieving this transformation in less than two minutes. This capability provides a 3325-fold increase in data resolution, which is particularly valuable for enhancing ultra-short-term wind forecasting applications.

The computational time required for a single IAS simulation averages 43 seconds, with EAS requiring 61 seconds. However, E-IAS simulations exceed 100 seconds due to the additional calculations needed for extrapolation and data loading. Nonetheless, this is still within our tolerable latency for the 15 minutes ahead periodic forecast context. Part of the stated time is the EAS, in case sufficient inlet measurements are present the IAS is run directly without the need for inlet extrapolation. The E-IAS model's ability to generate high-resolution wind field data makes it highly suitable for use in ultrashort-term forecasting models. By creating extrapolation look-up tables, the E-IAS framework can be seamlessly integrated into forecasting applications, improving accuracy without significantly increasing computational demands. Furthermore, statistical models can be easily updated following changes in terrain by reconstructing wind fields without requiring new measurement datasets.

While direct comparisons between different model types are generally discouraged, we evaluate WiSpEx's performance against state-of-the-art forecasting models from the literature to gain a comprehensive understanding of its capabilities. This approach aims to determine whether WiSpEx can be adapted and expanded for ultra-short-term wind power generation forecasting, with the intention of integrating it into WEEL. Since the core physics-based principles of WiSpEx will remain consistent for this new application, we do not expect significant changes in its behavior.

It is important to note that the positioning of the inlet relative to the WTs affects the phase accuracy of simulated wind speeds. Specifically, increasing the distance between the inlet and the WTs leads to greater phase errors due to the time required for an air parcel to travel this distance. Consequently, placing measurement instruments far from a WF suggests that WiSpEx's extrapolations could serve effectively as forecasts.

In Wang and Yang, 2021, the basic statistics of state-of-the-art forecasting models are presented, providing a baseline for roughly evaluating our model's performance, as shown below. At the first forecasting step, the MAPE of our model falls within the middle of the MAPE range of these models, which is promising given the simplifications implemented in our simulations. However, at the second forecasting step, the MAPE of all models significantly increases, ranging between 8.09% and 14.89%. Notably, the model proposed by Wang and Yang, 2021 almost doubles its MAPE to 8.09%. By the third forecasting step, their model establishes the lower limit for MAPE at 11.73%, further escalating to exceed 40%.

This behavior across all models highlights the challenges of predicting such a highly variable atmospheric condition over shorter time frames. Despite these challenges, our model demonstrates the potential to produce accurate estimations. The ability to generate 3D high-resolution datasets from 1D measurements with low errors is advantageous for any forecasting model, including those compared in the study. Creating reliable extrapolated wind datasets is critical for all stages of a WF, from selecting optimal locations for WTs to ensuring efficient operation of the WF and optimizing smart grid performance.

3.1.3 The W.E.E.L. Forecast Modelling Approach

Every WT extracts Energy from wind based on it's engineered design. The momentum of a WT temporarily delays it from being synchronized with its optimal rotational frequency for a given change in wind speed. This results in a power production latency(or lag) which we call Wind Energy Extraction Latency. Turbulence and strong gradients affect this latency. Our model, WEEL, named after the it's purpose, is created to unveil and exploit a WT's "power lag" to create, improve or extend (in time) Ultra-Short-Term Power Production forecasts through the conversion of wind speed to power by reducing the errors produced by this physical process in a fast and simple way.

WEEL captures the inherent lag between wind speed and power output with simple, fast, data driven statistical methods. Instead of relying on simple averaging of wind speed data, which can lead to a loss of valuable information regarding trends and short-term fluctuations, we propose utilizing the highest resolution Δt averaged measurements (available or interpolated) and Moving Average (MA)s instead of simple averaging. This approach helps maintain the original sample length while preserving key trend characteristics. More details are provided in [26].

The first part of WEEL is the Power Curve model. The Power Curve's approximation is the primary source of errors for the E-yield forecasting. The Power curves P(u) (u is the scalar velocity magnitude) can be produced by continuous functions approximated to real data, by averaging to speed bins (wind speed ranges), or be obtained by manufacturers. Most complex models include meteorological variables and account turbulence, focusing on how atmospheric conditions affect the power output of a WT as presented by [31]. Tailoring a curve to suit realistic needs is always important in any type and scope of modeling approach.

The second part of the model is the identification of the latency. The detection of latency is called Wind Optimal Shifted Value (ShiVa) identification. WEEL can directly produce ultra-high-time-frame forecasts quickly and directly by applying a power curve transformation to WEEL's maximum ShiVa shifted wind speed series. Maximum shiva is the maximum allowed shift in terms of error, while not raising the error more than the unshifted. WEEL uses the ShiVas to stabilize the error from power conversion and/or extend the forecasting horizon of a forecasting module. In our case, it utilizes WiSpEx, assuming that there is a phase error due to air travel distance from the inlet of the simulations to the WTs. Such a scenario is possible when measurement locations are far away from the WT vicinity, like in the case of conically scanning LIDARs.

The final conversion of wind speed to power is typically accomplished by generating a power curve from empirical data or using the manufacturer's specified power curve. The accuracy of the fitted power function and the standard deviation of the error significantly influence the performance of WEEL. To assess the performance of WEEL under realistic conditions, we consider a scenario where the function approximating the power curve of a WT captures the primary characteristics of power production but may not fully account for behavior at the extremes of the WT's operational range. For this purpose, we employ least squares optimization [35] in python to approximate the power curve with Pfit, which is based on the input variable x (e.g., wind speed) and several parameters.

Least squares optimization is a fundamental technique for fitting a model to data by minimizing the sum of squared differences between observed values and the values predicted by the model. The objective is to find the model parameters that make the predicted values as close as possible to the actual observations.

3.1.4 Experiments and Results

For our experiment, we focus on 15-minute power production forecasting due to its significance in providing WFs a reasonable time frame for operational adjustments and offering energy traders an opportunity to strategize. This focus defines the averaging window for power production. We employ a 10-minute averaging window for the wind speed MAs, matching the already extracted 10-min wind speed WiSpEx results. 10-minute MAs are both slightly faster and more volatile, a characteristic that can be particularly advantageous during ramp events.

Our validation process utilizes WTs' 10-minute MAs of wind speed measurements and 15-minute power measurements of 1-minute resolution to pinpoint the maximum ShiVa offset of wind speed series that minimizes the standard deviation with respect to their fitted power curve. By employing different averaging window

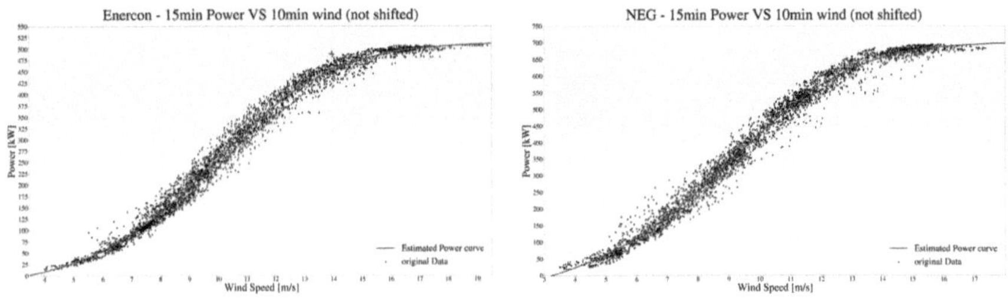

(a) (b)

Fig. 3.4 Enercon and NEG 15-min power curves from 0-shifted 10-min wind measurements. (**a**) Enercon—No shift. (**b**) NEG—No shift

sizes for wind speed and power calculations, we exploit the inherent trade-off between volatility and trend robustness. The 10-minute MA wind speed, while exhibiting higher volatility, captures trend changes more rapidly. Conversely, the 15-minute MA power output smooths out fluctuations but requires a slightly longer duration to reflect the underlying trend.

By identifying the WT "power lag" via visual scatter plot inspection for different time shifts (see Figs. 3.4, 3.5, and 3.6) and different error metrics (Tables 3.2, and 3.3), the volatility inherent in the wind speed data can be effectively mitigated. This is achieved by shifting the wind speed series forward by the determined optimal or maximum ShiVa. Once the optimal ShiVa is established, the estimation of 15-minute power production with a max ShiVa-minute lead time is possible, and the final power curve can be selected. This methodology proves particularly valuable for capturing power production during rapid changes in wind speed like ramp events, without incurring a significant computational burden.

In [26], we present and analyze the results obtained from WEEL model. Many ultra-short-term wind power forecasting models are being published the last few years, and almost all of them are statistical models like the ones presented by [22, 28, 38, 45]. Most of the models available present their results in terms of error reduction and not with direct error evaluations. Most of the times scientists use Symmetric Mean Absolute Percentage Error (SMAPE) instead of MAPE to reduce extreme error significance and normalized values like Normalized Root Mean Square Error (NRMSE) and Normalized Mean Absolute Error (NMAE) to generalize model behavior.

Table 3.4 provides a detailed overview of how well the WEEL method performs in the prediction of power output at different time steps for the Enercon WTs. A key takeaway is the consistently low SMAPE, which stays below 10% for all prediction horizons, indicating sufficient accuracy.

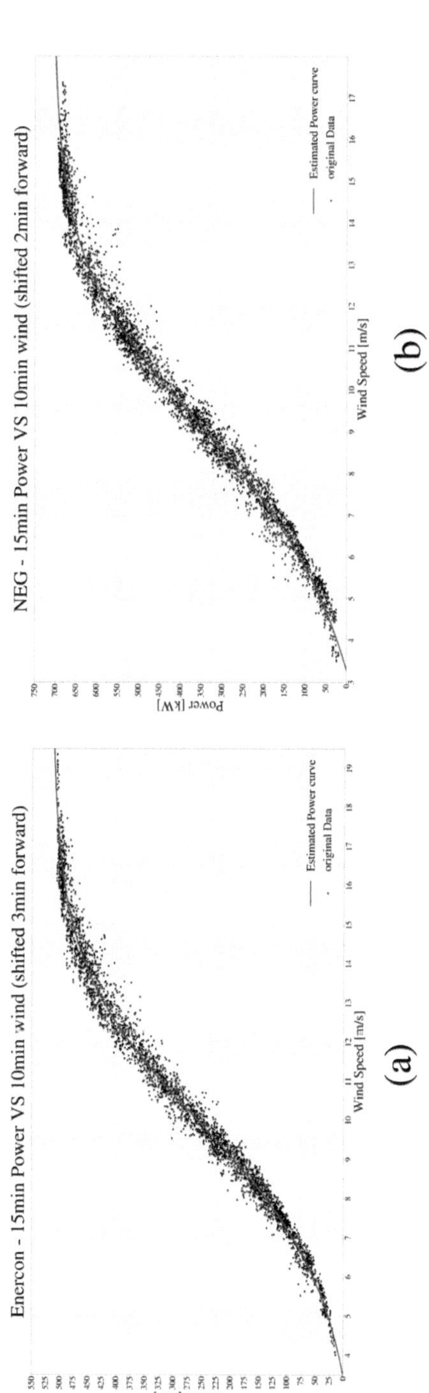

Fig. 3.5 Enercon and NEG 15-min power curves from 2–3 minutes shifted 10-min wind measurements. (**a**) Enercon—Optimal ShiVA shift. (**b**) NEG—Optimal ShiVA shift

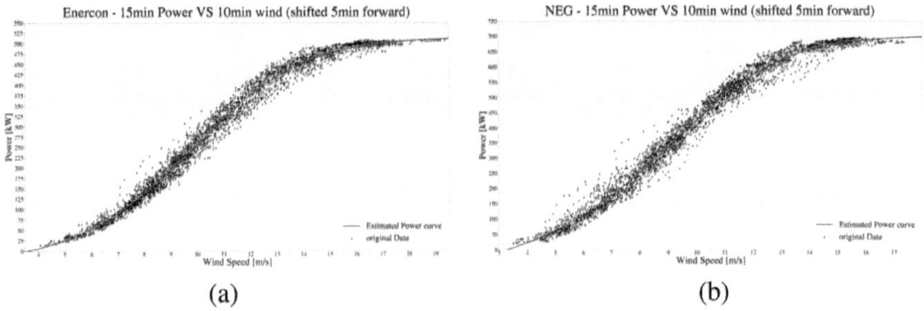

Fig. 3.6 Enercon and NEG 15-min power curves from 5-minute shifted 10-min wind measurements. (**a**) Enercon—Maximum ShiVA shift. (**b**) NEG—-Maximum ShiVA shift

Table 3.2 WEEL—Enercon shift stats

Metric	+0-min shift	+1-min shift	+2-min shift	+3-min shift	+4-min shift	+5-min shift
MAE	12.961	11.577	10.838	10.675	11.129	12.353
MSE	293.660	231.483	204.463	201.226	220.992	274.580
RMSE	17.136	15.214	14.299	14.185	14.865	16.570
MAPE	6.342	5.664	5.317	5.225	5.416	5.988

Table 3.3 WEEL—NEG shift stats

Metric	+0-min shift	+1-min shift	+2-min shift	+3-min shift	+4-min shift	+5-min shift
MAE	18.385	16.930	16.384	16.490	17.179	18.653
MSE	614.732	511.373	470.061	472.260	518.871	629.909
RMSE	24.793	22.613	21.680	21.731	22.778	25.097
MAPE	7.561	6.872	6.601	6.652	6.988	7.74

3.1.5 Discussion and Future Work

In our work we have demonstrated the complementary behavior between data-driven and physics-based approaches. It highlights the necessity of both approaches in a comprehensive wind power forecasting system. Data-driven models are invaluable for real-time operations and short-term decision-making, where capturing immediate fluctuations is critical. Physics-based models are widely used to provide valuable insights into longer-term trends and can serve as a reliable backup when historical data is scarce or unreliable. This is why the attempt to create a physics-based ultra-short-term forecasting model that has the ability to extrapolate data like WiSpEx is a difficult and valuable task. A significant advantage of physics-based model is their capability to directly adapt to any environment and changes that happen by human intervention. On the other hand, statistical models need training on new data for new locations and in cases where human interventions change the characteristics in the vicinity of a WT.

Table 3.4 Enercon 15-minute power forecast error stats for various forecast methods and time steps

Method	15-min MA	WEEL (Direct)	Persistence	WEEL Power Interpolated	Persistence	WEEL (Wi.Sp.Ex.)	Wi.Sp.Ex.	Persistence
MAE [kW]	5.06	12.23	15.32	12.96	24.85	17.81	31.58	31.82
MSE[kW^2]	45.18	271.15	450.05	294.81	1188.44	532.48	1643.36	1938.99
RMSE [kW]	6.72	16.47	21.21	17.17	34.47	23.08	40.54	44.03
MAPE [%]	2.4	5.91	7.41	6.1	12.2	8.44	15.36	16.01
NMAE	0.0101	0.0245	0.0306	0.0259	0.0497	0.0356	0.0632	0.0636
NRMSE	0.0134	0.0330	0.0424	0.0343	0.0690	0.0462	0.0811	0.0881
SMAPE [%]	2.39	5.88	7.29	5.99	11.72	8.18	14.51	15.03
Time step	live	+5 min	+5 min	+10 min	+10 min	+15 min	+15 min	+15 min

Recognizing the strengths and limitations of each approach, researchers are increasingly exploring hybrid models that combine data-driven (statistical) and physics-based components, like NWP forecasts [8, 9, 13, 14, 21, 33, 40, 44]. These hybrid models aim to leverage the short-term accuracy of data-driven techniques with the longer-term stability of physics-based approaches like NWP. By integrating both paradigms, they seek to achieve superior overall forecasting performance across various time horizons. Our fast modeling approach is an attempt to bridge the scale gap that is between mesoscale and microscale, especially in urban areas and their surroundings, as NWP models' horizontal resolution is usually 1x1 km or more (up to several kilometers). Ultimately, neither approach is universally superior. Their coexistence and potential integration into hybrid models hold the key to unlocking the full potential of wind power forecasting. As research in this field progresses, we can anticipate even more sophisticated models capable of accurately predicting wind power generation across various timescales, facilitating the seamless integration of this renewable resource into our energy systems.

While this study provides compelling evidence of WEEL's effectiveness, plenty of room for future research and development remain. Further investigations into WEEL's performance across diverse geographical locations and WF configurations are suggested. Analyzing its capabilities in varying terrains and wind regimes can yield deeper insights into its adaptability and potential limitations.

Exploring WEEL's application at the individual turbine level is another promising direction. Predicting the power output of each turbine separately and then aggregating the results could further reduce forecasting errors, especially in large WFs with heterogeneous wind conditions. Because our WiSpEx simulations are more accurate for higher WT tower heights, this favors our approach given the current technology trend.

Integrating WEEL with advanced remote sensing technologies, such as scanning long-range LIDARs, can unlock even greater levels of accuracy. Leveraging the high-resolution wind data captured by these technologies allows WEEL to refine its understanding of wind flow dynamics and generate more precise power predictions. Finally, investigating the potential of combining WEEL with other state of-the-art forecasting models, such as those based on machine learning or artificial intelligence, could lead to the development of hybrid forecasting approaches that harness the strengths of both physics-based and data-driven techniques.

3.2 Mid-Term Probabilistic Wind Power Forecasting

With wind power penetrating the energy systems more and more every year over the last decade, researchers have focused on quantifying the forecasting uncertainty. Especially for longer forecasting horizons, the intermittency and volatility of wind power can lead to forecasts of significant uncertainty, not captured by deterministic forecasting models which only offer single-valued predictions. On the other hand, wind power probabilistic forecasting (WPPF) models offer a broader range of potential outcomes, presenting results as quantiles, prediction intervals (PIs), or distributions. PIs, in particular, offer estimates about the upper and the lower bound of the generated outcome without compromising computational efficiency [4].

WPPF offers valuable insights into the fluctuations and risks associated with wind power generation, which facilitates more informed decision-making [5]. Consequently, probabilistic methods may eventually become as effective as deterministic ones, potentially transforming various decision-making processes, such as grid management, energy trading, and resource allocation, into probabilistic frameworks. For instance, energy traders can leverage probabilistic forecasts to refine their trading strategies, while grid operators can use this information to enhance energy supply management and ensure grid stability.

Various WPPF models have been proposed in recent years, focusing more on mid-term to long-term forecasting horizons, for which the forecasting uncertainty significantly increases. These models usually combine Numerical Weather Predictions (NWPs) and data-driven methods with probabilistic analysis to generate the final probabilistic forecasts. We refer to surveys [4] and [5] for a more detailed review of the current state of the art and trends of WPPF. In this section, we focus on data-driven methods for the forecasting uncertainty quantification, creating mid-term to long-term probabilistic forecasts in the form of PIs.

This section demonstrates the implementation of a mid-term WPPF model that generates intra-day and potentially even day-ahead PIs. Furthermore, it highlights the need to further enhance data-driven models with optimization algorithms, as well as to introduce hybrid forecasting and optimization approaches in order to achieve optimal probabilistic forecasting performance.

3.2.1 Case Studies

Two different case studies were investigated for the proposed WPPF approaches. The first dataset is the one featured in the Global Energy Forecasting Competition 2014 (GEFCom2014) [15]. It includes hourly data from 10 WFs in Australia, comprising wind observations at 2 different altitudes, 10 meters and 100 meters above ground level. For each of the 10 WFs, the dataset includes the zonal and meridional wind components (u10, v10, u100, v100) as well as the wind power output values normalized by the nominal capacity of each WF. The second dataset involves data from 200 randomly selected WTs within a WF and 3 meteorological masts. These WTs are situated in a flat area in the United States. The dataset provides hourly wind speed and power output data for each WT, along with hourly wind speed and direction measurements for each meteorological mast.

3.2.2 *Wind Power Probabilistic Forecasting Models*

A common issue in methodologies for constructing PIs from point predictions generated by data-driven models is the reliance on assumed data distributions. Although these distribution assumptions can streamline the PI construction process by simplifying forecasting error considerations, they can lead to complications if the actual data deviates from the assumed distribution. As a result, such methods are often not well suited for real-world applications and tend to be computationally intensive. Therefore, we developed two WPPF models based on the Lower Upper Bound Estimation (LUBE) method [6]. The key benefit of the LUBE method is its ability to simplify PI construction, as it only requires an ANN to directly estimate the lower and upper bounds of a PI. In order for the ANN to generate two point forecast time series, corresponding to the upper and the lower bounds of the PIs, it needs two neurons in the output layer that represent these bounds.

3.2.2.1 Singe-Objective Optimization

An improved Particle Swarm Optimization (PSO) algorithm was chosen for the calibration of the LUBE-based data-driven model. PSO is a metaheuristic optimization technique within the realm of swarm intelligence. Initially introduced as an evolutionary computation method, PSO addresses the optimization of both continuous and discontinuous functions in decision-making problems [17]. In PSO, each potential solution is represented by a particle within a swarm. During each iteration of the optimization process, particles explore the solution space, offering possible solutions to the problem at hand. The movement of the particles in the solution space is determined as follows: each particle is initially assigned a random position, representing a potential solution for that iteration of the optimization problem. The particle's movement is then influenced by its current movement direction, its best position found so far within the solution space, and the best position found so far by the swarm as a whole. A fitness function evaluates the position of each particle during each iteration, indicating how close it is to the optimal solution. Each particle memorizes its best position as p_{best}, and the best position obtained considering the whole swarm as g_{best}. The equations used to update the velocity and position of each particle in each iteration are the following:

$$v_n(t) = wv_n(t-1) + c_1 r_1(p_{best,n} - x_n(t-1)) + c_2 r_2(g_{best} - x_n(t-1)) \tag{3.1}$$

$$x_n(t) = x_n(t-1) + v_n(t-1) \tag{3.2}$$

where v_n and x_n represent the velocity and location of the nth particle at iteration t, $r_1, r_2 \epsilon [0, 1]$ are random variables, $c_1, c_1 \epsilon [1,2]$ are acceleration constants, and w is the inertia weight that is represented by

$$w = w_{max} - (w_{max} - w_{min}) \frac{t}{t_{max}} \tag{3.3}$$

where w_{max} and w_{min} define the initial and final inertia weights, respectively, and t_{max} is the maximum number of iterations. To enhance the global search capabilities of the particles and avoid local optima entrapment, their movement is further enhanced by an adaptive, Gaussian mutation operator.

The optimization of the final model is posed as a single-objective approach, by minimizing the Coverage Width Criterion (CWC):

$$CWC = \frac{1}{N} \sum_{i=1}^{N} (u_i - l_i) + \gamma e^{-h(\frac{1}{N} \sum_{i=1}^{N} (c_i) - cl)} \tag{3.4}$$

where γ is a binary parameter that equals 1 when the average coverage of the PIs is less than the nominal confidence level cl. The CWC achieves a trade-off between the accuracy and the sharpness of the PIs, as it aims to satisfy the pre-defined confidence level with the sharpest PIs possible.

In order to identify the optimal ANN structure for the proposed model, a k-fold cross-validation technique is employed. Feed-forward ANNs are utilized, with the number of neurons in the hidden layers varying from 1 to 20 in increments of 1 neuron. The k-fold cross-validation is applied to the training dataset to keep it distinct from the test set. In this method, the training set is divided into k complementary folds, with $k-1$ folds used for training the ANNs and the remaining fold for validation. This study implements a fivefold cross-validation process. PSO is then applied to optimize the ANN's weights and biases. Beyond determining the optimal ANN structure, a fivefold cross-validation is also conducted to establish the optimal number of particles in the PSO algorithm's swarm. Wavelet transformation was applied to decompose the wind power time series into simpler sub-series, and enhance data preprocessing. The provided wind power time series is decomposed into an approximation sub-series and four detail sub-series using a db4 mother wavelet function. The generated sub-series serve as inputs for the proposed LUBE-based ANN model, ultimately improving the model's overall accuracy.

3.2.2.2 Multi-Objective Optimization

Preserving the multi-objective nature of PI optimization (accuracy maximization—sharpness maximization) help reduce the overall induced bias by avoiding the artificial adjustments required to merge multiple contradicting objectives into one [23]. Furthermore, metaheuristic algorithms are generally more effective with less induced bias. Therefore, a multi-objective optimization framework was selected for the optimization of the second proposed WPPF model [20].

The multi-objective optimization framework is formulated as follows:

$$\begin{aligned}
&\text{maximize } \frac{1}{N}\sum_{t=1}^{N} c_t \\
&\text{minimize } \frac{1}{R}\sqrt{\frac{1}{N}\sum_{i=1}^{N}(u_i - l_i)^2} \\
&\text{subject to } \begin{cases} \frac{1}{N}\sum_{i=1}^{N} c_i \geqslant cl & \text{(Constraint 1)} \\ c_i = \begin{cases} 1 & \text{if } l_i \leqslant y_i \leqslant u_i \\ 0 & \text{else} \end{cases} & \text{(Constraint 2)} \\ 0 \leqslant l_i < u_i \leqslant 1 & \text{(Constraint 3)} \end{cases}
\end{aligned} \quad (3.5)$$

The problem objectives and constraints are outlined in 3.5, which focus on maximizing PI accuracy and sharpness. Specifically, the first objective maximizes the Prediction Interval Coverage Probability (PICP), which represents the average coverage of the PIs, while the second objective minimizes to the Prediction Interval Normalized Root Width (PINRW), which measures the normalized root of the average squared width of the PIs [16]. A smaller PI width indicates greater sharpness. The multi-objective problem formulation includes three constraints: the average coverage of the PIs must meet or exceed the pre-defined nominal confidence level. The parameter c_i is a binary variable indicating whether an actual observation falls within the corresponding PI. Additionally, PI bounds are constrained between 0 and 1, corresponding to zero and nominal wind power generation, respectively, while the upper and lower bounds of each PI must differ to prevent generating point predictions.

To effectively capture the temporal autocorrelation of the wind power time series, the feed-forward ANN of the first proposed model was replaced with a recurrent ANN. Due to the already increased complexity of the multi-objective approach, a nonlinear auto-regressive network with exogenous inputs (NARX) was selected. The NARX network is a straightforward data-driven model that effectively captures temporal dependencies in time-series data without compromising efficiency. This makes it a suitable choice for the proposed multi-objective optimization approach. The NARX network's core architecture resembles that of a feed-forward ANN, but it incorporates lagged output feedback into the input, along with other external inputs.

The architecture of the NARX network was determined as a result of a co-optimization process applied on the proposed hybrid multi-objective LUBE-based model. Extensive testing was conducted to determine the optimal structure of the NARX network. A second hidden layer was found to be necessary for modeling more complex features of WPPF inputs and outputs. Although adding more hidden layers can marginally enhance the forecasting performance in some cases, it also significantly increases computational complexity. To support the LUBE method with minimal computational overhead, feedback is provided by both outputs separately rather than combining them into a single value. This approach also reduces the induced bias in the model, which is advantageous for metaheuristics-based optimization. Feedback is given at two different lagged time steps, effectively capturing the temporal interdependencies in the wind power generation time series.

The employed Multi-Objective Improved Adaptive PSO (MOIAPSO) algorithm builds upon the classic multi-objective PSO described in [11], with enhancements through the use of an adaptive grid [10] and binary tournament selection for the selection of the swarm's leader. These enhancements are essential for balancing the trade-off between forecasting accuracy and training time. The adaptive grid promotes global exploration by increasing diversity among the stored non-dominated solutions, while the binary tournament method helps reduce overall training time. After evaluation, a repository is established to store non-dominated solutions, with the least diverse solutions being discarded in cases of repository overflow. The solutions in the repository are mapped according to their performance on each objective. An adaptive grid is then constructed to encompass all solutions and assess their diversity. Each hypercube of the grid is assigned a crowding index based on the number of solutions it contains, with a higher index indicating lower diversity. The grid dimensions are adjusted dynamically to accommodate all repository solutions. Binary tournament selection is employed to choose the swarm leader from the repository solutions, based on the crowding index. Mutation is applied to the new positions, which are updated using Eqs. 3.1 and 3.2. A particle's personal best position is updated if it is dominated by its current position; if the current position and the personal best position are nondominant relative to each other, the personal best is chosen randomly. Further information can be found in [20].

3.2.3 Results and Discussion

The proposed mid-term WPPF approaches are evaluated to estimate their efficiency and accuracy. The proposed single-objective optimization approach (LUBE-PSO-CWC) is compared with a Bootstrap Extreme Learning Machine (BELM) model, while the proposed multi-objective optimization approach (NARX-LUBE-MOIAPSO) is compared with a Chance Constrained Extreme Learning Machine (CCELM), a hybrid Long Short-Term Memory (LSTM)-Kernel Density Estimation (KDE) model, as well as a hybrid LSTM-LUBE model, which is optimized with the Non-dominated Sorting Genetic Algorithm II (NSGA II). All models were developed in Python, and experiments were conducted on an AMD Ryzen 5 3600X 6-Core Processor CPU (3.80GHz) desktop computer with 16 GB of RAM. The CWC and the Continuous Ranked Probability Score (CRPS) are used as probabilistic evaluation metrics. CRPS is calculated as follows:

$$CRPS = \frac{1}{N}\sum_{i=1}^{N}\int_{-\infty}^{\infty}(F(z_i) - l(z_i - y_i))^2\,dz \tag{3.6}$$

where F is the CDF of the generated forecasts and l is the Heaviside step function.

Table 3.5 Overall comparison of LUBE-PSO-CWC and BELM models

Comparisons	LUBE-PSO-CWC	BELM	Difference (%)
Average CWC	48.7056	51.2942	−5.05
Average CRPS	0.096859	0.10609	−9.53
Average run time (s)	126	27	366.67
Number of cases with best CWC	7	1	600.00
Number of cases with best CRPS	7	1	600.00

Table 3.6 Aggregated results of proposed NARX-LUBE-MOIAPSO model

Model	CRPS	CWC	Average training time
NARX-LUBE-MOIAPSO	48.7056	51.2942	−5.05
LSTM-LUBE-NSGA II	0.096859	0.10609	−9.53
LSTM-KDE	126	27	366.67
CCELM	7	1	600.00

In Table 3.5, an overall comparison between BELM and the proposed LUBE-PSO-CWC model is presented for the first case study (GEFCom 2014). The average CWC of LUBE-PSO-CWC is 0.487056, which is 5.05% less than the average CWC of BELM. The average PICP of LUBE-PSO-CWC is 0.929234, which is 2.29% more than the average PICP of BELM. In summary, both the average coverage rate (CWC) and the distribution of the forecasts (CRPS) are improved with the proposed model. As expected, however, BELM is faster, since its core consists of ELMs.

The aggregated results for the second case study are presented in Table 3.6. The proposed model clearly outperforms LSTM-KDE and CCELM for all evaluation metrics. LSTM-KDE and CCELM are the main competitors of the proposed model, due to their relative proximity regarding their computational complexity. Compared to LSTM-LUBE-NSGA II, the proposed model generates slightly worse results. Specifically, the proposed model's performance regarding CRPS is 6.7% worse than the performance of LSTM-LUBE-NSGA II. However, the average training time required for the proposed model is approximately 75% less than that required for LSTM-LUBE-NSGA II.

Figure 3.7 illustrates the PIs generated for 1 WT by the proposed NARX-LUBE-MOIAPSO model on a relatively "difficult" day, in which wind power fluctuations are frequent and intense. The larger the forecasting horizon, the higher the uncertainty of the forecast. This is mainly due to the inherited error of the meteorological parameter forecasts, which is propagated to the probabilistic WPF model. The proposed WPF model captures the wind power time-series behavior sufficiently up to about an 8-hour-ahead forecasting horizon. Hence, the proposed model is suitable for forecasts from one to several hours ahead (intra-day forecasting).

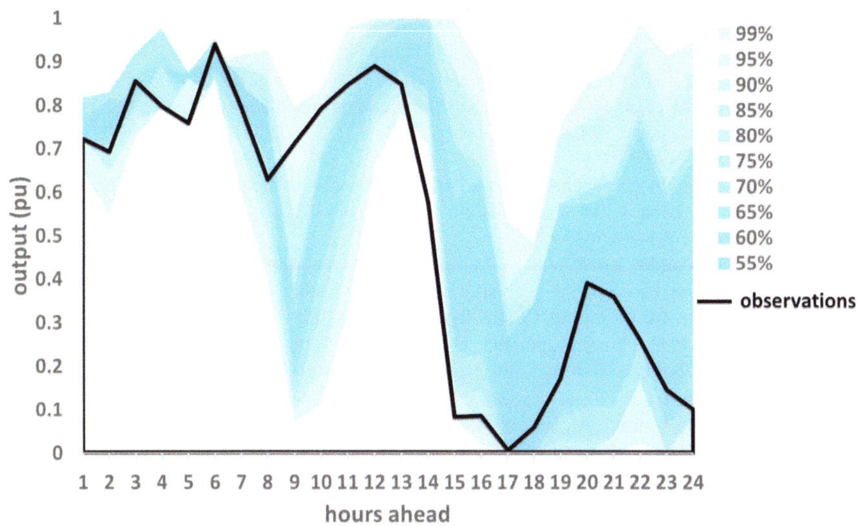

Fig. 3.7 24-hour-ahead PIs generated for different confidence levels by the proposed NARX-LUBE-MOIAPSO model, as well as the real power output of the corresponding WT

References

1. Anagnostos, D., Schmidt, T., Goverde, H., Kalisch, J., Catthoor, F., Soudris, D.: Pv energy yield nowcasting combining sky imaging with simulation models. In: 31st European Photovoltaic Solar Energy Conference and Exhibition, pp. 1552–1555 (2015)
2. Baıle, R., Muzy, J.F.: Leveraging data from nearby stations to improve short-term wind speed forecasts. Energy **263**, 125–644 (2023). https://doi.org/10.1016/j.energy.2022.125644
3. Barbieri, F., Rajakaruna, S., Ghosh, A.: Very short-term photovoltaic power forecasting with cloud modeling: A review. Renew. Sustain. Energy Rev. **75**, 242–263 (2017)
4. Bazionis, I., Karafotis, P., Georgilakis, P.: A review of short-term wind power probabilistic forecasting and a taxonomy focused on input data. IET Renew. Power Gener. **16** (2021)
5. Bazionis, I.K., Georgilakis, P.S.: Review of deterministic and probabilistic wind power forecasting: models, methods, and future research. Electricity **2**(1), 13–47 (2021)
6. Bazionis, I.K., Kousounadis-Knudsen, M.A., Konstantinou, T., Georgilakis, P.S.: A WT-LUBE-PSO-CWC wind power probabilistic forecasting model for prediction interval construction and seasonality analysis. Energies **14**(18) (2021)
7. Bingöl, F., Mann, J., Foussekis, D.: 2.3 lidar in complex terrain. Advancements in Wind Energy Metrology–UPWIND 1A2. 3 p. 11 (2011). https://orbit.dtu.dk/en/publications/advancements-in-wind-energy-metrology-upwind-1a23
8. Chang, Y., Yang, H., Chen, Y., Zhou, M., Yang, H., Wang, Y., Zhang, Y.: A hybrid model for long-term wind power forecasting utilizing nwp subsequence correction and multi-scale deep learning regression methods. IEEE Trans. Sustain. Energy **15**(1), 263–275 (2024). https://doi.org/10.1109/TSTE.2023.3283242
9. Chen, N., Qian, Z., Nabney, I.T., Meng, X.: Wind power forecasts using gaussian processes and numerical weather prediction. IEEE Trans. Power Syst. **29**(2), 656–665 (2014). https://doi.org/10.1109/TPWRS.2013.2282366
10. Coello, C., Pulido, G., Lechuga, M.: Handling multiple objectives with particle swarm optimization. IEEE Trans. Evol. Comput. **8**(3), 256–279 (2004)

11. Coello Coello, C., Lechuga, M.: Mopso: a proposal for multiple objective particle swarm optimization. In: Proceedings of the 2002 Congress on Evolutionary Computation. CEC'02 (Cat. No.02TH8600), vol. 2, pp. 1051–1056 (2002)
12. Dai, X., Liu, G.P., Hu, W.: An online-learning-enabled self-attention-based model for ultra-short-term wind power forecasting. Energy **272**, 127173 (2023). https://doi.org/10.1016/j.energy.2023.127173
13. Du, P.: Ensemble machine learning-based wind forecasting to combine NWP output with data from weather station. IEEE Trans. Sustain. Energy **10**(4), 2133–2141 (2019). https://doi.org/10.1109/TSTE.2018.2880615
14. He, B., Ye, L., Pei, M., Lu, P., Dai, B., Li, Z., Wang, K.: A combined model for short-term wind power forecasting based on the analysis of numerical weather prediction data. Energy Rep. **8**, 929–939 (2022). https://doi.org/10.1016/j.egyr.2021.10.102
15. Hong, T., Pinson, P., Fan, S., Zareipour, H., Troccoli, A., Hyndman, R.J.: Probabilistic energy forecasting: global energy forecasting competition 2014 and beyond. Int. J. Forecasting **32**, 896–913 (2016)
16. Kabir, H.M.D., Khosravi, A., Hosen, M.A., Nahavandi, S.: Neural network-based uncertainty quantification: a survey of methodologies and applications. IEEE Access **6**, 36218–36234 (2018)
17. Kennedy, J., Eberhart, R.: Particle swarm optimization. In: Proceedings of ICNN'95—International Conference on Neural Networks, vol. 4, pp. 1942–1948 (1995)
18. Kim, D., Kim, T., Oh, G., Huh, J., Ko, K.: A comparison of ground-based lidar and met mast wind measurements for wind resource assessment over various terrain conditions. J. Wind Eng. Ind. Aerodyn. **158**, 109–121 (2016). https://doi.org/10.1016/j.jweia.2016.09.011
19. Kogaki, T., Sakurai, K., Shimada, S., Kawabata, H., Otake, Y., Kondo, K., Fujita, E.: Field measurements of wind characteristics using lidar on a wind farm with downwind turbines installed in a complex terrain region. Energies **13**(19), 5135 (2020). https://doi.org/10.3390/en13195135
20. Kousounadis-Knousen, M.A., Bazionis, I.K., Soudris, D., Catthoor, F., Georgilakis, P.S.: A new co-optimized hybrid model based on multi-objective optimization for probabilistic wind power forecasting in a spatio–temporal framework. IEEE Access **11**, 84885–84899 (2023)
21. Liu, C., Zhang, X., Mei, S., Zhou, Q., Fan, H.: Series-wise attention network for wind power forecasting considering temporal lag of numerical weather prediction. Appl. Energy **336**, 120815 (2023). https://doi.org/10.1016/j.apenergy.2023.120815
22. Liu, L., Liu, J., Ye, Y., Liu, H., Chen, K., Li, D., Dong, X., Sun, M.: Ultra-short-term wind power forecasting based on deep bayesian model with uncertainty. Renew. Energy **205**, 598–607 (2023). https://doi.org/10.1016/j.renene.2023.01.038
23. Lu, P., Ye, L., Zhao, Y., Dai, B., Pei, M., Tang, Y.: Review of meta-heuristic algorithms for wind power prediction: Methodologies, applications and challenges. Appl. Energy **301**, 117446 (2021)
24. Manfren, M., Caputo, P., Costa, G.: Paradigm shift in urban energy systems through distributed generation: methods and models. Appl. Energy **88**(4), 1032–1048 (2011). https://doi.org/10.1016/j.apenergy.2010.10.018
25. Michos, D., Catthoor, F., Foussekis, D., Kazantzidis, A.: A CFD model for spatial extrapolation of wind field over complex terrain—wI. Sp. Ex. Energies **17**(16), 4139 (2024). https://doi.org/10.3390/en17164139
26. Michos, D., Catthoor, F., Foussekis, D., Kazantzidis, A.: Ultra-short-term wind power forecasting in complex terrain: a physics-based approach. Energies **18** (2025). Accepted for publication
27. Nallolla, C.A., P, V., Chittathuru, D., Padmanaban, S.: Multi-objective optimization algorithms for a hybrid ac/dc microgrid using res: a comprehensive review. Electronics **12**(4), 1062 (2023). https://doi.org/10.3390/electronics12041062
28. Niu, D., Sun, L., Yu, M., Wang, K.: Point and interval forecasting of ultra-short-term wind power based on a data-driven method and hybrid deep learning model. Energy **254**, 124384 (2022). https://doi.org/10.1016/j.energy.2022.124384

29. Polimeni, S., Nespoli, A., Leva, S., Valenti, G., Manzolini, G.: Implementation of different PV forecast approaches in a multigood microgrid: modeling and experimental results. Processes **9**(2), 323 (2021). https://doi.org/10.3390/pr9020323
30. Quiñones, J.J., Pineda, L.R., Ostanek, J., Castillo, L.: Towards smart energy management for community microgrids: leveraging deep learning in probabilistic forecasting of renewable energy sources. Energy Conver. Manag. **293**, 117440 (2023). https://doi.org/10.1016/j.enconman.2023.117440
31. Saint-Drenan, Y.M., Besseau, R., Jansen, M., Staffell, I., Troccoli, A., Dubus, L., Schmidt, J., Gruber, K., Simões, S.G., Heier, S.: A parametric model for wind turbine power curves incorporating environmental conditions. Renew. Energy **157**, 754–768 (2020). https://doi.org/10.1016/j.renene.2020.04.123
32. Sezer-Uzol, N., Long, L.: 3-d time-accurate CFD simulations of wind turbine rotor flow fields. Am. Inst. Aeron. Astron. **394** (2006). https://doi.org/10.2514/6.2006-394
33. Shirzadi, N., Nasiri, F., Menon, R.P., Monsalvete, P., Kaifel, A., Eicker, U.: Smart urban wind power forecasting: Integrating weibull distribution, recurrent neural networks, and numerical weather prediction. Energies **16**(17) (2023). https://doi.org/10.3390/en16176208
34. Valldecabres, L., Peña, A., Courtney, M., von Bremen, L., Kühn, M.: Very short-term forecast of near-coastal flow using scanning lidars. Wind Energy Sci. **3**(1), 313–327 (2018). https://doi.org/10.5194/wes-3-313-2018
35. Virtanen, P., Gommers, R., Oliphant, T.E., Haberland, M., Reddy, T., Cournapeau, D., Burovski, E., Peterson, P., Weckesser, W., Bright, J., van der Walt, S.J., Brett, M., Wilson, J., Millman, K.J., Aldcroft, T., Mayorov, N., Jones, E., Smith, R., Colbert, S., et al.: SciPy 1.0: fundamental algorithms for scientific computing in Python. Nat. Methods **17**(3), 261–272 (2020). https://doi.org/10.1038/s41592-019-0686-2
36. Wei, J., Wu, X., Yang, T., Jiao, R.: Ultra-short-term forecasting of wind power based on multi-task learning and LSTM. Int. J. Electr. Power Energy Syst. **149**, 109073 (2023). https://doi.org/10.1016/j.ijepes.2023.109073
37. Wu, Y.T., Porté-Agel, F.: Large-eddy simulation of wind-turbine wakes: evaluation of turbine parametrisations. Boundary-Layer Meteorol. **138**(3), 345–366 (2011). https://doi.org/10.1007/s10546-010-9569-x
38. Xiang, L., Liu, J., Yang, X., Hu, A., Su, H.: Ultra-short term wind power prediction applying a novel model named SATCN-LSTM. Energy Conver. Manag. **252**, 115036 (2022). https://doi.org/10.1016/j.enconman.2021.115036
39. Xiaoxia, G., Luqing, L., Shaohai, Z., Xiaoxun, Z., Haiying, S., Hongxing, Y., Yu, W., Hao, L.: Lidar-based observation and derivation of large-scale wind turbine's wake expansion model downstream of a hill. Energy **259**, 125051 (2022). https://doi.org/10.1016/j.energy.2022.125051
40. Yakoub, G., Mathew, S., Leal, J.: Intelligent estimation of wind farm performance with direct and indirect 'point' forecasting approaches integrating several NWP models. Energy **263**, 125893 (2023). https://doi.org/10.1016/j.energy.2022.125893
41. Yan, B., Li, Q.: Coupled on-site measurement/CFD based approach for high-resolution wind resource assessment over complex terrains. Energy Conver. Manag. **117**, 351–366 (2016). https://doi.org/10.1016/j.enconman.2016.02.076
42. Yan, S., Shi, S., Chen, X., Wang, X., Mao, L., Liu, X.: Numerical simulations of flow interactions between steep hill terrain and large scale wind turbine. Energy **151**, 740–747 (2018). https://doi.org/10.1016/j.energy.2017.12.075
43. Yang, B., Zhong, L., Wang, J., Shu, H., Zhang, X., Yu, T., Sun, L.: State-of-the-art one-stop handbook on wind forecasting technologies: an overview of classifications, methodologies, and analysis. J. Cleaner Prod. **283**, 124628 (2021). https://doi.org/10.1016/j.jclepro.2020.124628
44. Yang, M., Wang, D., Zhang, W.: A short-term wind power prediction method based on dynamic and static feature fusion mining. Energy **280**, 128226 (2023). https://doi.org/10.1016/j.energy.2023.128226

45. Zhang, Y., Han, J., Pan, G., Xu, Y., Wang, F.: A multi-stage predicting methodology based on data decomposition and error correction for ultra-short-term wind energy prediction. J. Clean. Prod. **292**, 125981 (2021). https://doi.org/10.1016/j.jclepro.2021.125981
46. Zhou, L., Hu, G., Tse, K.T., He, X.: Twisted-wind effect on the flow field of tall building. J. Wind Eng. Ind. Aerodyn. **218**, 104778 (2021). https://doi.org/10.1016/j.jweia.2021.104778

Chapter 4
Online HVAC Control for Energy Efficiency and Thermal Comfort

Charalampos Marantos, Christos Sad, Kostas Siozios, and Dimitrios Soudris

Building facilities are increasingly becoming active participants in the evolving energy market. The growing demand for energy efficiency, combined with rapid advancements in renewable energy and smart-grid IoT technologies, is driving the need for autonomous control solutions, particularly for energy-intensive building subsystems such as heating, ventilation, and air-conditioning (HVAC). HVAC systems play a crucial role in maintaining optimal indoor environments but consume significant amounts of energy. In addition, utilities and information, such as market-driven pricing, are becoming increasingly accessible to end users. This allows consumers to make more informed decisions about building energy consumption and encourages the adoption of more efficient energy practices by promoting energy use during low-price periods and discouraging heavy consumption during peak times.

The ongoing developments in smart grids, energy markets, and IoT technologies present both opportunities and challenges. On one hand, smart HVAC control systems can now be designed to automatically adjust their operation based on real-time conditions such as occupancy, weather forecasts, and energy prices, maximizing thermal comfort for occupants. On the other hand, the continuous demand to reduce energy consumption and carbon footprints, along with the dynamic nature of energy pricing, has introduced a new priority for these systems: energy efficiency. For example, when renewable sources are more abundant, the HVAC system can be programmed to operate more intensively, ensuring the building is adequately cooled or heated. Conversely, when energy is purchased for the main grid and prices peak,

C. Marantos · D. Soudris
Zografou Campus, National Technical University of Athens, Athens, Greece

C. Sad · K. Siozios (✉)
Aristotle University of Thessaloniki, Thessaloniki, Greece
e-mail: ksiop@auth.gr

F. Catthoor et al. (eds.), *Energy Production, Load and Battery Management Framework with Supporting Methods for Smart Microgrids*,
https://doi.org/10.1007/978-3-031-92025-7_4

the system can be scaled back to reduce costs without significantly compromising thermal comfort.

Such intelligent control mechanisms are becoming increasingly important as buildings transition from passive energy consumers to active participants in the energy market. These solutions not only enhance the sustainability of building operations but also reduce operational costs and carbon footprints, contributing to the development of green, smart cities. This shift toward smarter, more responsive HVAC systems represents a major advancement in building energy management [3]. In this context, this chapter presents a low-complexity online HVAC control solution that targets existing buildings and optimizes both energy consumption and thermal Comfort, without requiring detailed building models. The presented solution not only exhibits negligible complexity enabling its implementation on low-cost embedded devices (e.g., smart thermostats) without affecting the quality of the derived results but also offers shorter design times, alleviating the time-to-market pressure.

4.1 Problem Definition

This section introduces the template of our case study and the problem definition. The study targets a micro-grid environment encompassing multiple energy sources (i.e., purchasing energy from the main grid and installed PV systems), as well as the HVAC systems, which are the primary energy consumers. The focus is on the orchestration task: determining the HVAC thermostat set-points.

To support the HVAC control task, various installed sensors collect data related to weather (temperature, humidity, solar radiation, and forecasts), building conditions (indoor temperature and humidity), and residents' activities. This data is transmitted to the main controller, which computes optimal actions to co-optimize thermal comfort and energy cost metrics.

Before proceeding to the details of the introduced orchestration task, we must discuss the analytical form of the targeted HVAC operation optimization problem. We formulate this as a weighted-sum multi-objective optimization problem (MOO) (Eq. 4.1). The two objectives—energy consumption (E) and occupants' thermal comfort (C)—are minimized and ω is the weight/trade-off between them. It's worth mentioning here that thermal comfort (C) is quantified as the *Predicted Percentage of Dissatisfied* occupants. Also, s_i denotes the input variables corresponding to the temperature set-points of the HVAC system at timestep i. Finally, the objective functions are influenced by an external vector of environmental variables, α_i.

$$Min\{\omega \times E(s_t, \alpha_t) + (1 - \omega) \times C\,(s_t, \alpha_t)\} \tag{4.1}$$

Providing an accurate analytical description and a plug&play solution at the same time is challenging due to the need for complex prior modeling for each new/existing building. Consequently, the problem addressed in this chapter has the following characteristics: (i) the objective functions depend not only on the temperature set-point but also on the building's environment and occupants' behavior while their detailed form is unknown; (ii) the controller actions can be evaluated discretely at each timestep and no prior information is given to the solver.

By considering these properties, the chapter presents an online plug&play solution that effectively manages HVAC operations in a dynamic and data-driven manner (without any prior information), ensuring both energy efficiency and occupant comfort optimization.

4.2 Related Work

The problem addressed in this chapter is a well-known challenge that has gained significant attention from researchers over the years [18, 23, 25]. Traditionally, the most accurate solutions have relied on detailed modeling of buildings and HVAC systems, often applying control theory methods or model predictive control (MPC) [4]. While these methods can achieve high accuracy, they are complex and resource-intensive, both in terms of design time and operational requirements. The accurate modeling of all the dynamics of a building is a very challenging task, making such approaches mostly applicable to new constructions. Furthermore, the computational requirements of these control algorithms, which typically involve the solution of non-convex, high-dimensional mathematical problems [5], make their implementation feasible only to enterprise environments, where the necessary computational resources are available.

However, recent advancements in smart IoT systems, coupled with the growing need for solutions that can be applied to existing buildings [12], including residential ones, have led to the development of smart thermostats and smart stand-alone, building-agnostic air-conditioners. These solutions are designed to be plug&play, generic, and cost-effective. Unlike traditional methods, they often rely on online data analysis and algorithms that adapt in real time to changing conditions.

Several approaches have been explored in this domain, including genetic algorithms [8], empirical models [24], simulation optimization [7], event-based optimization (EBO) [23], fuzzy logic [2, 9, 17], and artificial neural networks (ANNs) [10]. These methods, while effective in the operation phase, often have limited flexibility when applied to existing buildings. They typically require a pretraining phase based on historical data, which can take years to collect, leading to long design times and creating solutions that are tightly bound to a specific building, making them unsuitable as a plug&play option for existing structures. Additionally, these approaches impose significant computational and storage complexities, particularly during the training phase.

A modern approach that has gained considerable attention in recent years is reinforcement learning (RL). RL is one of the most promising methods for providing a "black-box" and plug&play solution [1, 22]. In RL-based solutions, the optimization problem is modeled to consist of a set of states, a set of actions, and a reward function. At each timestep, the agent (the HVAC system's controller) samples the current state of the building and computes the next thermostat set-point (action), with the goal of maximizing the received reward (optimizing energy consumption and thermal comfort) [13, 14].

However, RL-based solutions come with a set of challenges. One major issue is the stability—RL systems can take a significant amount of time to stabilize, which can be problematic in dynamic environments. Additionally, when external factors such as weather conditions change or when parameters of the reward function, such as the weight (ω) in Eq. 4.1, are modified, the system may need to be retrained from scratch. Finally, another limitation is the risk of the agent getting trapped in local minima during the optimization process, necessitating sophisticated exploration mechanisms.

4.3 Decision-Making Mechanism

The introduced solution aims to improve and mitigate the limitations of using the aforementioned "black-box" approaches as well as to be flexible and able of supporting multiple operating scenarios.

The enabler of the presented approach is the introduction of simple (low complexity) models that achieve improved accuracy based on only a small subset of the received sensors data.

4.3.1 Preprocessing Steps

First of all, the system collects the data from the sensors and builds tuples of (*Outdoor temperature* (T_t^{out}), *Solar radiation* (R_t), *Indoor temperature* (T_t^{in}), *Indoor humidity* (H_t)), connected with an Energy consumption (E_t) and a calculated Thermal Comfort (C_t) for each timestep (t) to build the dataset that will support the control task (Eq. 4.2):

$$\alpha_t = \{[T_t^{out}, R_t, T_t^{in}, H_t], E_t, C_t\} \tag{4.2}$$

Regarding the energy cost (E_t), if the expected energy load of the building at timestep t exceeds the energy available from the photovoltaic panels (PVs), the additional demand is met by purchasing energy from the main grid at the prevailing price. Conversely, if the energy requirements are within the capacity of the microgrid's renewable sources, the building's energy needs are fully met by these sources, eliminating the need for additional grid energy.

Then, the cost is calculated based on Eq. 4.1, as a weighted sum of the two objectives (E_t and C_t), while the weighting factor ($\omega \in [0, 1]$) gives the relative importance between them.

4.3.2 Selective Historical Data Management

The key enabler of the proposed online method is the efficient management of historical data collected by the building's sensors. Rather than storing the entire raw data, this approach selectively retains data corresponding to specific days or hours, optimizing both storage and computational resources.

To achieve this, the method employs two complementary sliding window mechanisms, the coarse-grain and fine-grain windows, as illustrated in Fig. 4.1. The coarse-grain window retains data from the last x days, while the fine-grain window focuses on storing data from specific periods within each day [15].

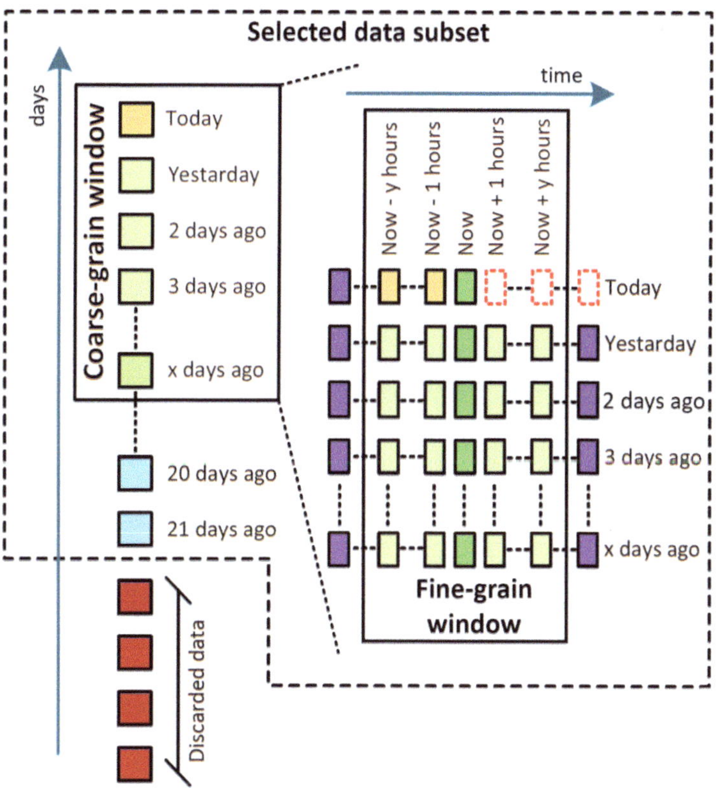

Fig. 4.1 Efficient historical data selection

This selective data storage strategy significantly reduces the amount of data that needs to be stored, making it feasible to use simple and low-complexity methods, such as linear regression, for supporting the HVAC control. By using the proposed sliding windows, the online HVAC control method not only conserves storage space but also accelerates the learning process, allowing for fast adaptation based on a limited amount of data. This model-free solution is particularly well-suited for deployment on low-cost embedded devices commonly found in smart thermostats.

In the following sections, we will present a wide range of experiments that demonstrate the effectiveness of this method. These experiments highlight how this approach contributes to the development of fast yet accurate models for estimating energy consumption and thermal comfort across different buildings, all without the need for extensive prior modeling.

4.3.3 Core Decision-Support Algorithm

The core decision-support process for online HVAC control is illustrated in Fig. 4.2. This process begins with the collection of data from various sources, including

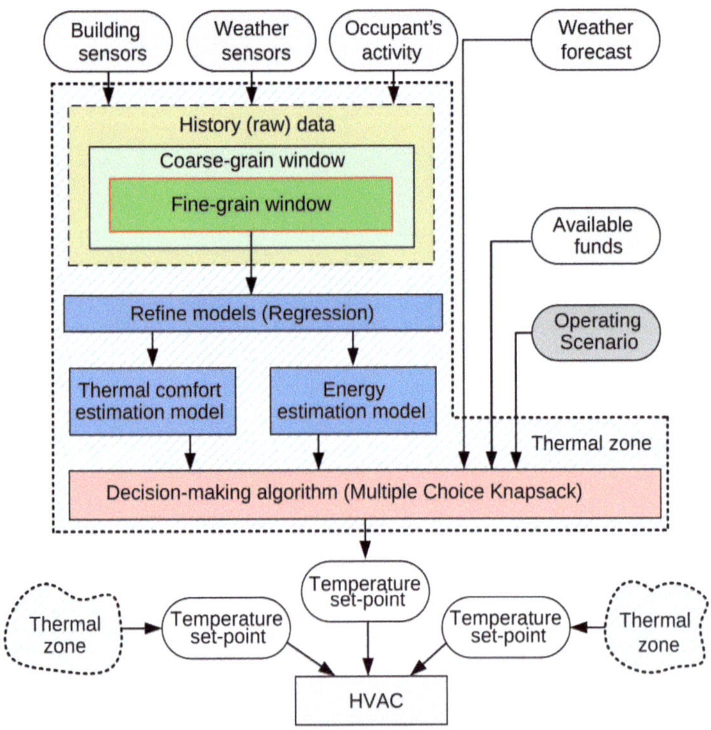

Fig. 4.2 Core decision-support algorithm

building sensors, occupant activity, weather forecasts, and user input regarding available funds and the desired operational scenario. This data is then efficiently managed and stored using the selective data management mechanism described earlier.

Once the relevant data is selected, it is used to refine two key regression models: one for predicting energy consumption and another for forecasting thermal comfort. These models are then fed into the final decision-making step, which employs a Multiple-Choice Knapsack algorithm. This algorithm is responsible for determining the optimal HVAC thermostat set-points according to the specified operating scenario [16].

The final stage of this process is the decision-making step, which involves the application of a Multiple-Choice Knapsack algorithm. This algorithm is responsible for calculating the optimal HVAC set-points based on the given operating scenario. The Multiple-Choice Knapsack Problem (MCKP) is defined as follows: a set of items is divided into distinct groups, and exactly one item from each group must be selected to be included in the knapsack. The goal is to maximize the total value of the selected items, while ensuring that the sum of their weights does not exceed the knapsack's capacity. However, in our application, the objective is inverted as we aim to minimize total cost rather than maximize value and thus each item is assigned a negative value (representing cost).

The algorithm supports three distinct operating scenarios:

- Scenario 1: Optimize Both Energy Consumption and Thermal Comfort
  In this scenario, the weights of the items are set to zero, and the negative value (cost) is defined as the weighted sum of energy consumption and the predicted percentage of dissatisfied (PPD) occupants in terms of thermal comfort, as described in Eq. 4.1.
- Scenario 2: Optimize Energy Consumption Without Violating Thermal Comfort Constraints
  Here, the negative values (cost) correspond to the energy consumption of each option, while the weights represent the PPD. The objective is to minimize energy consumption while adhering to thermal comfort limits.
- Scenario 3: Optimize Thermal Comfort Without Exceeding the available funds for Energy
  In this scenario, the negative values (cost) are derived from the PPD values, and the weights correspond to the energy consumption of each choice. The knapsack's capacity is defined by the available funds for energy, ensuring that thermal comfort is maximized without surpassing the budget.

By adapting the Multiple-Choice Knapsack algorithm to these specific scenarios, the decision-support system ensures that the HVAC control strategy aligns with the user's priorities, whether they focus on energy efficiency, thermal comfort, or a balanced approach to both.

4.4 Energy and Thermal Comfort Estimation

As discussed earlier, efficient data management using the two complementary sliding window mechanisms is the key enabler for employing simple models, such as linear regression or decision trees, for energy consumption and thermal comfort estimation. In this section, we will present the design, implementation, and results of this approach.

4.4.1 Preparing the Data

At each timestep, data collected from the building sensors are stored in the system's memory, utilizing the coarse-grain sliding window mechanism presented in Sect. 4.3.2. This approach ensures that data older than the coarse-grain window size are discarded in a First-In-First-Out (FIFO) manner. The format of the stored data is detailed in Table 4.1. Prior to applying the energy and thermal comfort estimation models, the fine-grain sliding window mechanism is employed. This process selectively forwards only a subset of the stored data, corresponding to specific hours of the day, to the estimators.

4.4.2 Energy Estimation

To evaluate the effectiveness of the proposed energy estimation solution, we applied a linear regression approach on the data retrieved by the proposed sliding window selection to predict the energy consumption for the next hour in three different types of buildings. The first test was conducted using a simulation in the widely accepted EnergyPlus software.[1] This approach offers several advantages, such as the ability to modify temperature set points, HVAC configurations, location, weather conditions, and occupant activity, allowing for the testing of various scenarios. However, to address the possibility that the simulation and the building models created for the experiments may not fully replicate real-world conditions, we also evaluated the

Table 4.1 Stored data collected by the sensors

$$\begin{bmatrix} T_1^{out} & R_1 & T_1^{in} & H_1 \\ T_2^{out} & R_2 & T_2^{in} & H_2 \\ \vdots & \vdots & \vdots & \vdots \\ T_m^{out} & R_m & T_m^{in} & H_m \end{bmatrix} \quad \begin{bmatrix} E_1 \\ E_2 \\ \vdots \\ E_m \end{bmatrix} \quad \begin{bmatrix} C_1 \\ C_2 \\ \vdots \\ C_m \end{bmatrix}$$

[1] https://energyplus.net/

proposed approach using real-world residential building data (a subset of residential buildings) available online [11]. It is important to note that this energy data includes not only HVAC energy consumption but also the total energy consumption of the building, including other appliances.

The EnergyPlus building model is an office building, located in Crete Greece, operating from 9am to 6pm every day. The results for a representative week of the year are shown in Fig. 4.3. The accuracy is more than 95% for the entire year.

When it comes to residential buildings, energy estimation becomes more challenging. Unlike companies and universities, residential occupants do not typically follow a strict schedule, and the use of appliances, HVAC configurations, and other factors can be more unpredictable. However, as shown in Fig. 4.4, the energy estimation for the next hour remains acceptable, achieving up to 86.9% accuracy, a Spearman correlation between the actual and the predicted values of 0.87, and an R2-score of 0.69, largely because it is closely related to the energy consumption of the previous hour.

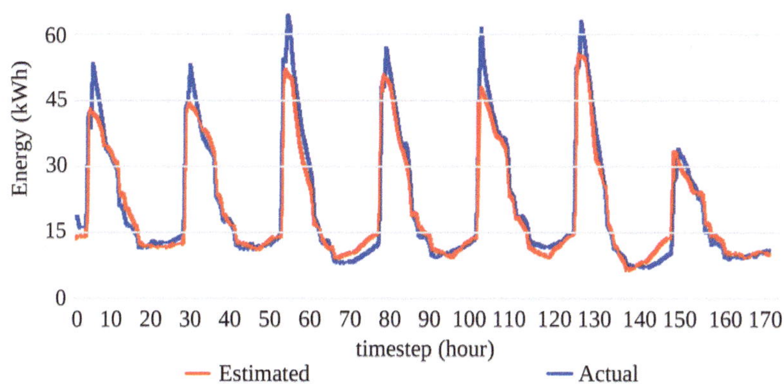

Fig. 4.3 Energy estimation for the office building simulated in EnergyPlus

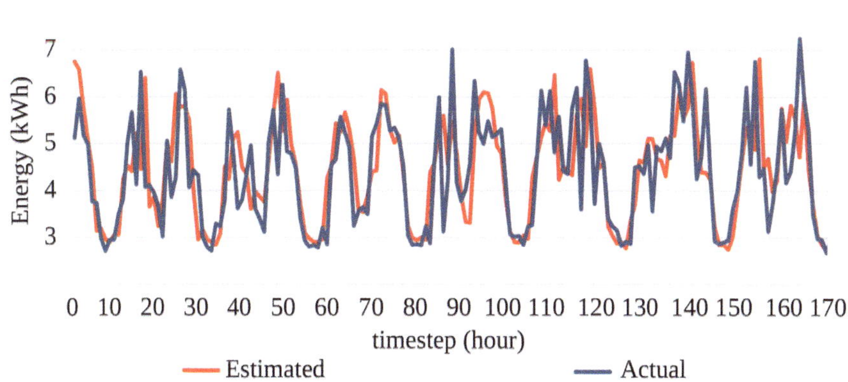

Fig. 4.4 Energy estimation for residential buildings

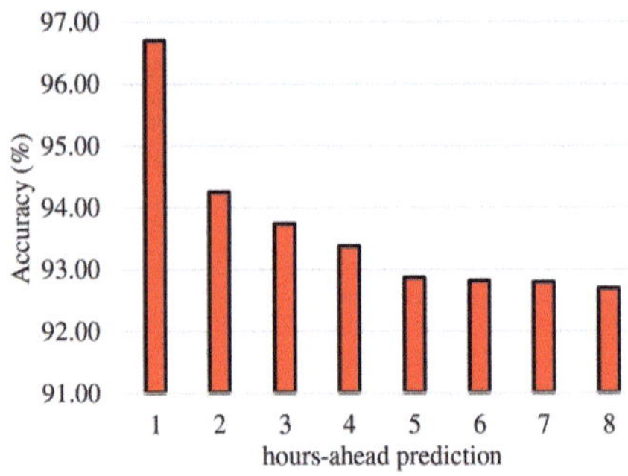

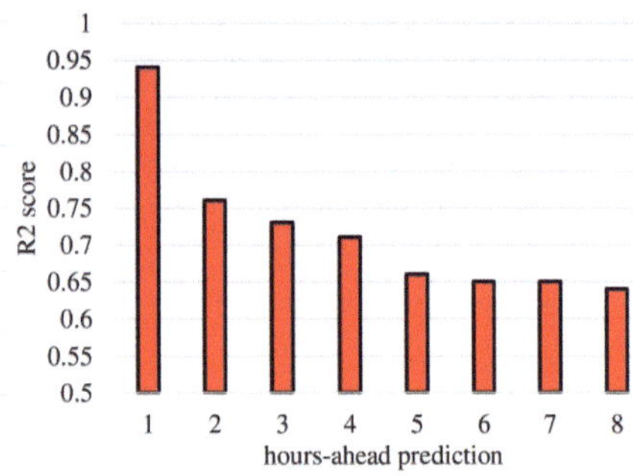

Fig. 4.5 Accuracy decrease for future hours energy estimation

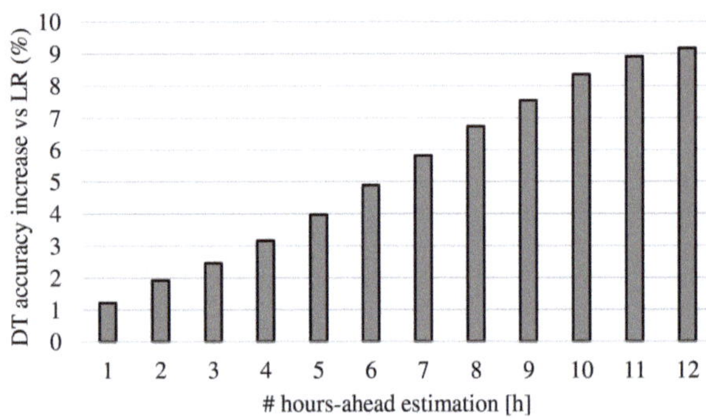

Fig. 4.6 Estimation accuracy increase for using decision tree instead of linear regression

Predicting energy consumption for several hours ahead proves more difficult, and the accuracy of predictions decreases. We observe that decision tree exhibits increased accuracy for future hours estimations [19], while linear regression remains the best for estimating the energy of the next hour. Figures 4.5 and 4.6 illustrate the estimation accuracy for the next 8 hours and the percentage increase of accuracy for using decision-tree estimators instead of linear regression respectively. Despite the observed reduction in accuracy, these results are still within an acceptable range and provide valuable inputs for the proposed HVAC control algorithm, as will be demonstrated in the rest of this chapter.

4.4.3 Thermal Comfort Model

The thermal comfort estimation model is based on the observation that it follows a quadratic function, as illustrated schematically in Fig. 4.7. Specifically, the HVAC configuration influences only the temperature, which in turn affects thermal comfort. Figure 4.7 demonstrates the Predicted Percentage of Dissatisfied (PPD) values

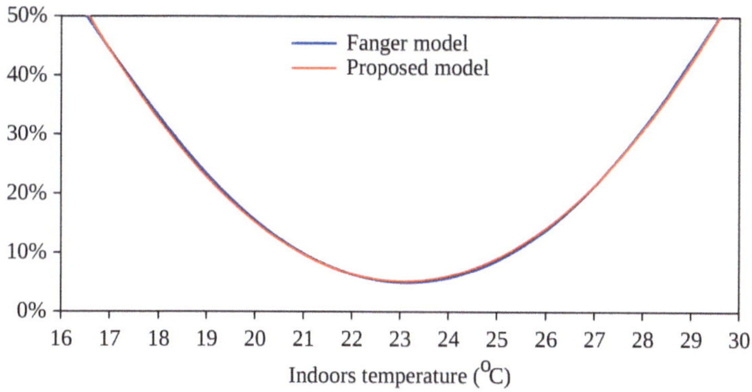

Fig. 4.7 PPD estimation with the proposed and the reference Fanger [6] models as a function of indoor temperature

for thermal comfort, according to Fanger's model [6], while keeping all other factors (such as occupants' activity, clothing, etc.) constant and varying only the temperature. Consequently, the proposed thermal comfort estimation is expressed in Eq. 4.3, by refining the θ_1 and θ_2 weights [16, 20, 21]. To achieve this, the HVAC controller solves the minimization problem outlined in Eq. 4.4. For thermal comfort estimation, the dataset used as input to this minimization, along with the associated weights θ_c, is provided in Eq. 4.5.

$$C_{est} = \theta_0^c + \theta_1^c \times T^{in} + \theta_2^c \times T^{in^2} \tag{4.3}$$

$$Min\{J_c(\theta^c))\}, \text{ where } J_c(\theta^c) = \sum\left(C_{real} - C_{est}\right)^2 = \left\|C_{real} - X_c \times \theta^c\right\|^2 \tag{4.4}$$

$$\theta^c = \begin{bmatrix} \theta_0^c \\ \theta_1^c \\ \theta_2^c \end{bmatrix}, \; X_c = \begin{bmatrix} 1 \; T_1^{in} \; (T_1^{in})^2 \\ 1 \; T_2^{in} \; (T_2^{in})^2 \\ \vdots \;\; \vdots \;\;\; \vdots \\ 1 \; T_m^{in} \; (T_m^{in})^2 \end{bmatrix}, \; C_{real} = \begin{bmatrix} C_1 \\ C_2 \\ \vdots \\ C_m \end{bmatrix} \tag{4.5}$$

4.4.4 *Impact of Selective Historical Data on Models Accuracy*

The impact of the historical data management strategies described in Sect. 4.3.2 is illustrated in Fig. 4.8. Based on these results, we have chosen a coarse-grain window size of 15 days and a fine-grain window size of 3 hours. This selection effectively minimizes the root mean square error (RMSE) for both energy and comfort estimations. Specifically, with the coarse-grain window, we achieve an

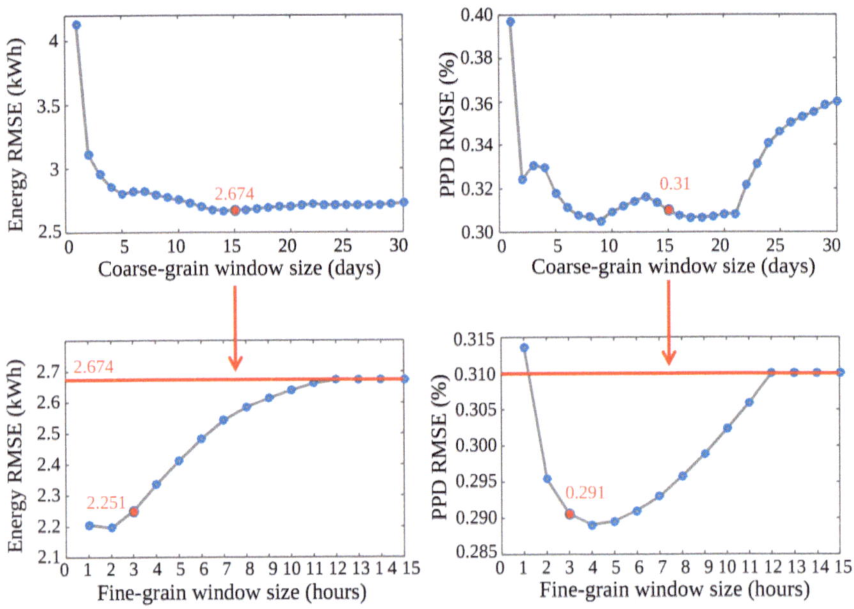

Fig. 4.8 RMSE analysis for quantifying the impact of the data selection window sizes to the energy estimation accuracy and the thermal comfort estimation accuracy

RMSE of 2.674 kWh for energy and 0.31% for the Predicted Percentage of Dissatisfied (PPD) occupants. The incorporation of the fine-grain window further reduces the RMSE, resulting in values of 2.251 kWh for energy and 0.291% for PPD.

4.5 Experimental Results

4.5.1 Experimental Setup

To evaluate the proposed algorithm, static data retrieved from buildings is insufficient, as we need to assess the performance of the HVAC configurations generated by the control algorithm in a dynamic environment. Therefore, a simulator is required. For this purpose, we employed EnergyPlus, a high-quality building simulation software, widely used by researchers, practitioners, and industry professionals.

Simulating an online algorithm, such as the one proposed, requires the ability to adjust thermostat set points in real time during the experiment. To facilitate this, we utilized a complementary program that communicates with EnergyPlus—namely, the Building Controls Virtual Test Bed (BCVTB). BCVTB allows users to interface with MATLAB, enabling real-time adjustments of building parameters.

The simulated buildings used in this study are real office buildings located in Crete, Greece, and were designed as part of the PEBBLE FP7 EU project.

4.5.2 Operating Scenarios: Comparison Against Fixed Set-Points

The first scenario focuses on minimizing a balanced cost that equally considers both energy consumption and thermal comfort (with the weight factor ω in Eq. 4.1 set to 0.5). The results, compared against maintaining constant thermostat set points at a fixed temperature configuration, are presented for a representative winter and summer day in Fig. 4.9. These results demonstrate that the proposed controller performs very well, achieving significant improvements in both energy efficiency and thermal comfort.

The second scenario aims to optimize energy consumption while ensuring that the Predicted Percentage of Dissatisfied (PPD) occupants does not exceed a certain threshold (with ω in Eq. 4.1 set to 1). According to the ASHRAE standard, a PPD level below 10% is considered acceptable for occupants. Based on the results for this scenario, shown in Fig. 4.10, the proposed control solution achieves to a 48% reduction in energy consumption compared to the fixed thermostat configurations that satisfy the thermal comfort standards.

The third scenario addresses optimizing thermal comfort without exceeding the available energy budget (funds). For the demonstration results presented in Fig. 4.11, we selected two different budget levels per week: a high-funds case and a low-funds case. These selections were not entirely random as they took into account the weather conditions and the average consumption at fixed temperature settings. In the high-funds scenario, we observe that the PPD level approaches 5%, which, according to the thermal comfort model, is considered the optimal score, as there

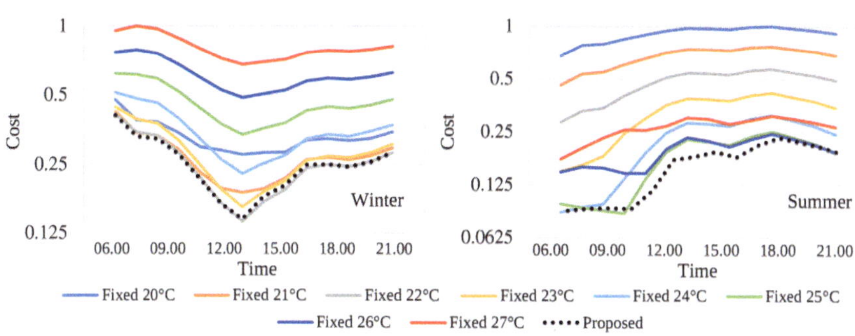

Fig. 4.9 Proposed controller efficiency against fixed thermostat values (balanced scenario)

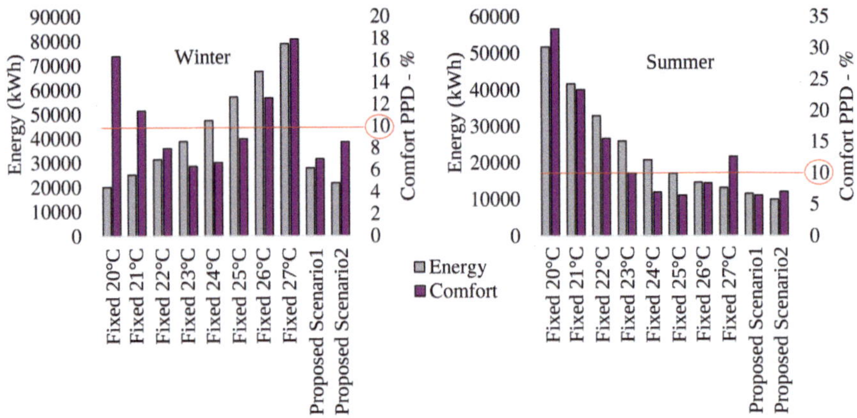

Fig. 4.10 Energy and PPD evaluation for the Scenario 2 vs fixed values

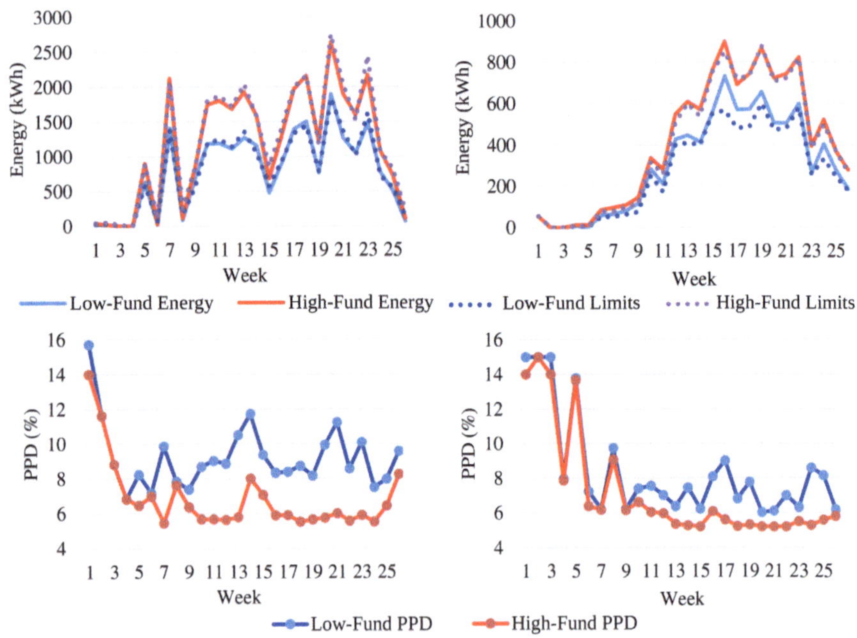

Fig. 4.11 Evaluate energy and PPD variation for alternative Scenario 3

will always be a small percentage (around 5%) of occupants who are dissatisfied due to individual preferences. In the low-funds case, while PPD increases, it remains within standard limits. Additionally, the proposed solution adheres to the budget constraints for all weeks, except in rare instances within the low-funds scenario where it was not possible to meet the PPD standard limit, without exceeding the very low budget.

4.5.3 Comparison Against Relative Approaches

In this section, we compare the proposed approach with several related solutions. Specifically, the comparison is made in the balanced scenario (Scenario 1), where energy consumption and thermal comfort are given equal importance ($\omega = 0.5$ in Eq. 4.1). The results are based on a yearlong experiment and are summarized in Table 4.2. The solutions compared are as follows:

- Fixed Thermostat Configurations: This approach involves maintaining a constant thermostat setting (e.g., fixed temperature of 20 °C), without any adjustments.
- Fmincon (Simulation-Based): This solution relies on repeated simulations of a modeled building. For our evaluation, the EnergyPlus building models were used, and results were obtained after 3000–5000 iterations. In this comparison, Fmincon is considered the most accurate solution, as it uses the same building models employed in our experiments. In a real-world scenario, these results would be achievable if perfectly accurate building models were available.
- Reinforcement Learning [13, 14]: This solutions use a "black-box" approach based on reinforcement learning, optimizing HVAC control without explicit modeling of the building or the objectives.
- SVM-Min [15]: This is a preliminary version of our proposed solution, combining sliding window data management with support vector machine (SVM) estimators.

Based on these results, we conclude that the proposed approach delivers superior performance. Specifically, it outperforms fixed thermostat configurations by 40% to 150%, is 5.5% better than the Reinforcement Learning approach, and is only 3% less effective than the best possible results achieved with a 100% accurate building model (Fmincon). Moreover, the proposed solution is highly efficient in terms of computational complexity, making it suitable for implementation on low-

Table 4.2 Balanced scenario annual results comparison against other methods

Method	Energy Consumption	Avg. PPD	Norm. Cost
Fixed 20 °C	66967 kWh	24.9%	0.89
Fixed 21 °C	62939 kWh	17.4%	0.72
Fixed 22 °C	61223 kWh	11.6%	0.59
Fixed 23 °C	61955 kWh	7.9%	0.52
Fixed 24 °C	65191 kWh	6.4%	0.51
Fixed 25 °C	70467 kWh	7.2%	0.56
Fixed 26 °C	77359 kWh	10.2%	0.66
Fixed 27 °C	85680 kWh	15.3%	0.81
Fmincon (Simulation-based)	34936 kWh	6.1%	0.33
Reinforcement Learning [13, 14]	34601 kWh	7.7%	0.36
SVM-Min [15]	36767 kWh	6.5%	0.35
Proposed	36399 kWh	6.4%	0.34

end smart devices, such as smart thermostats. Notably, the execution time is 8 orders of magnitude lower than that of the Fmincon MPC solution, requiring only 0.004 secs on an ARM Cortex-A57 for Scenarios 1 and 2. Even for the more demanding Scenario 3, which involves evaluating one week ahead and solving a more complex Knapsack problem at each timestep, the execution time does not exceed 0.03 seconds.

4.6 Conclusions

A framework for designing a low-cost orchestrator targeting HVAC systems was introduced, aimed at solving the multi-objective problem of simultaneously improving building energy consumption and occupant's thermal comfort.

The problem formulation, the objectives and the proposed decision-making algorithm, as detailed in this chapter, were designed to be flexible, supporting multiple operational modes/scenarios. These modes include balancing energy consumption with resident thermal comfort, minimizing energy use while maintaining acceptable thermal comfort levels, and maximizing thermal comfort without exceeding the available energy budget. The newly introduced models for estimating thermal comfort and HVAC energy consumption are both fast and accurate, exhibiting negligible complexity compared to similar implementations, without any loss in quality.

Our experimentation validates the superiority of the proposed solution against relevant solvers, eliminating the need for accurate prior system modeling, as both functions describing energy consumption and thermal comfort are agnostic. The presented orchestrator demonstrates efficiency comparable to the initial offline solver, with only a 3% reduction in total cost savings but significantly lower complexity—approximately 8 orders of magnitude less—without relying on prior knowledge or system modeling. Consequently, the introduced solutions are well-suited for low-cost embedded devices, such as smart thermostats, and enable shorter design times, thus reducing time-to-market pressure.

References

1. Barrett, E., Linder, S.: Autonomous HVAC control, a reinforcement learning approach. In: Joint European Conference on Machine Learning and Knowledge Discovery in Databases, pp. 3–19. Springer, Berlin (2015)
2. Calvino, F., La Gennusa, M., Rizzo, G., Scaccianoce, G.: The control of indoor thermal comfort conditions: introducing a fuzzy adaptive controller. Energy Build. **36**(2), 97–102 (2004)
3. Charalampos, M.: Design methodologies and tools for energy-aware IoT-based applications. Ph.D. Thesis, National Technical University of Athens (2023)

4. Clarke, J., Conner, S., Fujii, G., Geros, V., Jóhannesson, G., Johnstone, C., Karatasou, S., Kim, J., Santamouris, M., Strachan, P.: The role of simulation in support of internet-based energy services. Energy Build. **36**(8), 837–846 (2004)
5. Dounis, A.I., Caraiscos, C.: Advanced control systems engineering for energy and comfort management in a building environment—a review. Renew. Sustain. Energy Rev. **13**(6-7), 1246–1261 (2009)
6. Fanger, P.O.: Thermal Comfort: Analysis and Applications in Environmental Engineering. Danish Technical Press (1970)
7. Fong, K.F., Hanby, V.I., Chow, T.T.: HVAC system optimization for energy management by evolutionary programming. Energy Build. **38**(3), 220–231 (2006)
8. Huang, W., Lam, H.: Using genetic algorithms to optimize controller parameters for HVAC systems. Energy Build. **26**(3), 277–282 (1997)
9. Kolokotsa, D., Stavrakakis, G., Kalaitzakis, K., Agoris, D.: Genetic algorithms optimized fuzzy controller for the indoor environmental management in buildings implemented using PLC and local operating networks. Eng. Appl. Artif. Intell. **15**(5), 417–428 (2002)
10. Kumar, R., Aggarwal, R., Sharma, J.: Energy analysis of a building using artificial neural network: a review. Energy Build. **65**, 352–358 (2013)
11. Makonin, S.: HUE: the hourly usage of energy dataset for buildings in British Columbia. Data Brief **23**(103744), 103744 (2019)
12. Marantos, C., Lamprakos, C., Siozios, K., Soudris, D.: Towards plug&play smart thermostats for building's heating/cooling control. In: IoT for Smart Grids: Design Challenges and Paradigms, pp. 183–207 (2019)
13. Marantos, C., Lamprakos, C., Siozios, K., Soudris, D.: Model-free HVAC optimizer based on reinforcement learning. In: 2023 IEEE 32nd International Symposium on Industrial Electronics (ISIE), pp. 1–8. IEEE, Piscataway (2023)
14. Marantos, C., Lamprakos, C.P., Tsoutsouras, V., Siozios, K., Soudris, D.: Towards plug&play smart thermostats inspired by reinforcement learning. In: Proceedings of the Workshop on Intelligent Embedded Systems Architectures and Applications, pp. 39–44 (2018)
15. Marantos, C., Siozios, K., Soudris, D.: A flexible decision-making mechanism targeting smart thermostats. IEEE Embed. Syst. Lett. **9**(4), 105–108 (2017)
16. Marantos, C., Siozios, K., Soudris, D.: Rapid prototyping of low-complexity orchestrator targeting cyberphysical systems: the smart-thermostat usecase. IEEE Trans. Control Syst. Technol. **28**(5), 1831–1845 (2019)
17. Singh, J., Singh, N., Sharma, J.: Fuzzy modeling and control of HVAC systems—a review. J. Sci. Ind. Res. **65**, 470–476 (2006)
18. Vaghefi, S.A., Jafari, M.A., Zhu, J., Brouwer, J., Lu, Y.: A hybrid physics-based and data driven approach to optimal control of building cooling/heating systems. IEEE Trans. Autom. Sci. Eng. **13**(2), 600–610 (2014)
19. Watari, D., Marantos, C., Taniguchi, I., Catthoor, F., Siozios, K., Soudris, D., Onoye, T.: Online energy management framework for smart buildings with low-complexity estimators. IEEE Embed. Syst. Lett. 16(2), 138–141 (2024)
20. Watari, D., Taniguchi, I., Catthoor, F., Marantos, C., Siozios, K., Shirazi, E., Soudris, D., Onoye, T.: Thermal comfort aware online energy management framework for a smart residential building. In: 2021 Design, Automation & Test in Europe Conference & Exhibition (DATE), pp. 535–538. IEEE, Piscataway (2021)
21. Watari, D., Taniguchi, I., Catthoor, F., Marantos, C., Siozios, K., Shirazi, E., Soudris, D., Onoye, T.: Thermal-comfort aware online co-scheduling framework for HVAC, battery systems, and appliances in smart buildings. IEICE Trans. Fund. Electron. Commun. Comput. Sci. **106**(5), 698–706 (2023)
22. Wei, T., Wang, Y., Zhu, Q.: Deep reinforcement learning for building HVAC control. In: Proceedings of the 54th annual design automation conference 2017, pp. 1–6 (2017)
23. Wu, Z., Jia, Q.S., Guan, X.: Optimal control of multiroom HVAC system: an event-based approach. IEEE Trans. Control Syst. Technol. **24**(2), 662–669 (2015)

24. Yao, Y., Lian, Z., Hou, Z., Zhou, X.: Optimal operation of a large cooling system based on an empirical model. Appl. Thermal Eng. **24**(16), 2303–2321 (2004)
25. Zhang, X., Shi, W., Li, X., Yan, B., Malkawi, A., Li, N.: Decentralized temperature control via HVAC systems in energy efficient buildings: an approximate solution procedure. In: 2016 IEEE Global Conference on Signal and Information Processing (GlobalSIP), pp. 936–940. IEEE, Piscataway (2016)

Chapter 5
Multi-Timescale Energy Management Framework

Daichi Watari, Dafang Zhao, Ittetsu Taniguchi, and Francky Catthoor

In real-time energy management, fluctuations in renewable generation, such as photovoltaic (PV) systems, can result in significant energy losses and imbalances. Conversely, at slower time scales, scheduling distributed energy resources (DERs), such as battery systems and shiftable appliances, is crucial for shifting energy demand from peak to off-peak hours to reduce electricity costs. To bridge the gap between fast and slow time scales in energy management, this chapter introduces an online multi-time scale energy management framework for smart PV systems.[1] The proposed framework incorporates both coarse- and fine-grained time scales and iteratively solves three sequential optimization problems at predetermined time intervals. This multi-time scale approach reduces computational complexity while preserving solution quality. Furthermore, a short-to-middle term PV and hybrid equivalent-circuit battery model are employed for precise energy management. This framework enhances both modeling accuracy and computational efficiency in real-time energy management systems (EMSs). Results demonstrate that the proposed framework can reduce electricity costs by up to 47.5% compared to baseline methods while maintaining reasonable computational times.

[1] This chapter is a refined and reproduced version of the paper to be published in Applied Energy [42], copyrighted by Elsevier.

D. Watari · D. Zhao (✉) · I. Taniguchi
University of Osaka, Osaka, Japan
e-mail: zhao.dafang@ist.osaka-u.ac.jp

F. Catthoor
Zografou Campus, National Technical University of Athens, Athens, Greece

F. Catthoor et al. (eds.), *Energy Production, Load and Battery Management Framework with Supporting Methods for Smart Microgrids*,
https://doi.org/10.1007/978-3-031-92025-7_5

5.1 Motivation and Objective

With increasing environmental concerns, renewable energy sources, particularly solar and wind power, have been integrated into demand-side energy systems to reduce both CO_2 emissions and electricity costs [40]. However, renewable energy generation is inherently variable and difficult to control, with output fluctuating due to environmental factors such as cloud cover and wind speed. These variations can lead to supply-demand mismatches [24]. Smart energy systems incorporating renewable energy, battery storage, and demand control mechanisms offer potential solutions for improving energy efficiency and system resilience [27].

EMSs play a central role in smart energy systems by coordinating energy flow among appliances, batteries, renewable generation, and grid purchases [7]. These systems typically aim to optimize multiple objectives: reducing electricity costs, managing peak demand, and improving renewable energy utilization [28]. However, the variability of renewable sources can impact EMS performance [30]. Current systems often utilize battery storage operating at fast time scales to address these fluctuations [29], highlighting the importance of accurate battery modeling and appropriate integration of workload-dependent energy storage [5, 37].

Model predictive control (MPC) has shown promise in EMS applications for managing renewable energy variability and system dynamics. MPC enables systematic adjustment of control inputs by predicting and optimizing system behavior over defined time horizons. This approach has been applied to battery scheduling in conjunction with renewable energy forecasting to help manage uncertainties in smart energy systems [44].

Demand load control represents another important aspect of energy management [38]. Appliance scheduling, which coordinates load patterns with renewable generation and electricity prices [3, 26], allows flexible appliances to operate within specified time windows. While this approach can reduce costs and improve PV self-consumption, it operates on a slower time scale than that required for managing rapid PV fluctuations. This temporal disparity introduces additional complexity to energy management.

A fundamental challenge in smart energy systems lies in the inherent multiplicity of operational time scales, which poses significant computational and control challenges. These systems must simultaneously address two distinct temporal domains: rapid fluctuations occurring at the second level and longer-term variations spanning hours to days [12]. Specifically, while appliance scheduling decisions are typically made at daily or hourly intervals to optimize energy consumption patterns [36], the management of renewable energy fluctuations demands rapid, second-level control responses to maintain system stability. The significant difference between these time scales creates substantial computational challenges when attempting to integrate both control requirements into a single optimization framework. The challenge is particularly acute because the optimization objectives and system dynamics at each time scale are fundamentally different, yet inherently coupled through their effects on overall system performance.

This chapter proposes a multi-time scale energy management framework for smart PV systems. The framework is designed for local energy networks that integrate PV generation, battery storage, and controllable loads across multiple buildings. The proposed approach implements MPC with two key components: (1) PV forecasting based on neural networks and thermal models [2] and (2) an equivalent circuit battery model that incorporates state-of-charge dynamics and I-V characteristics [8]. To address the computational challenges while maintaining control performance, the framework employs a two-time scale structure with coarse-grained and fine-grained time scales. This structure enhances both computational efficiency and modeling capabilities, ensuring that the system can respond effectively to varying dynamics.

The main contributions of this work are as follows:

- Development of a two-time scale framework that systematically addresses both fast dynamics (second-level PV fluctuations and battery transient responses) and slow dynamics (hourly/daily appliance scheduling and demand profile management), enabling effective energy management across multiple temporal resolutions.
- Integration of three detailed component models for precise energy management: a time-shiftable appliance model incorporating user preferences and operational constraints, a physics-based PV forecasting model providing second-level resolution predictions, and an equivalent circuit battery model that captures nonlinear charge-discharge characteristics and state-dependent behavior.
- Implementation of MPC that simultaneously manages PV forecasting uncertainty and battery state evolution while maintaining computational tractability. The controller systematically coordinates decisions across both time scales to achieve efficient real-time operation.
- Comprehensive quantitative evaluation through simulation studies using measured PV and demand data, examining system performance under various PV forecasting errors (12–40%), different battery capacities (3–18 kWh), and multiple operational scenarios. The analysis demonstrates both computational feasibility and control effectiveness across different temporal resolutions.

The remaining of this chapter is organized as follows: Sect. 5.2 discusses related work on energy management methodologies for smart PV systems. Section 5.3 provides an overview of the proposed multi-time scale framework and the system models employed. Section 5.4 presents a detailed mathematical formulation of the energy management problem within the proposed framework. Section 5.5 demonstrates the effectiveness of the proposed method through extensive simulations using measured data. Finally, Sect. 5.6 summarizes the key findings and contributions of this chapter.

5.2 Related Studies

EMSs have been extensively studied using various control methodologies to optimize energy consumption and cost efficiency. Traditional approaches have focused on optimization-based scheduling for managing appliances and battery systems. Lokeshgupta et al. [25] proposed a home energy management system (HEMS) utilizing multi-objective mixed-integer programming (MIP) to simultaneously minimize electricity costs and peak power demands while evaluating battery investment economics for residential applications. Building on this approach, Dorahaki et al. [10] developed an advanced EMS using mixed-integer nonlinear programming (MINLP) that incorporates both electrical and thermal demand control alongside day-ahead energy storage scheduling. Their sensitivity analysis demonstrated the importance of coordinating electrical demand and battery utilization. However, these optimization-based approaches typically execute scheduling only once per day, lacking mechanisms to handle PV forecast errors or adapt to real-time changes in battery system states.

To address these limitations, researchers have increasingly adopted MPC strategies. Godina et al. [16] demonstrated MPC's superiority over traditional ON/OFF and PID controllers for air-conditioning management under dynamic pricing, achieving significant improvements in both cost reduction and control effectiveness. This success led to broader applications, with Parisio et al. [32] extending MPC to multi-building scenarios by incorporating weather forecasts, demand predictions, and user preferences. Carli et al. [6] further validated MPC's practical feasibility in real buildings, while Gan et al. [14] integrated renewable generation forecasting with MIP-based MPC for battery scheduling. These implementations typically utilize planning horizons of several days with coarse-grained resolutions (15–60 minutes), focusing primarily on slow dynamics management.

The challenges of real-time control at shorter time scales have led to alternative approaches. Rule-based controllers [43] and fuzzy logic systems [4] are commonly employed due to their computational efficiency. However, these methods cannot guarantee optimal solutions due to their reliance on simplified models and current-state information.

Recognizing the need to address multiple time scales simultaneously, researchers have explored hierarchical control structures. Abreu et al. [1] developed a two-layer MPC system for demand load management, while Lefort et al. [23] proposed a hierarchical EMS combining long-horizon scheduling with short-term operational control. Jin et al. [19] integrated day-ahead scheduling with intra-hour MPC adjustments for electric vehicle optimization with PV generation. Despite these advances, these approaches offer limited demand flexibility and employ simplified PV forecasting models.

A notable attempt to bridge these gaps was made by Elkazaz et al. [11], who developed a two-layer home EMS combining 24-hour scheduling with real-time control. However, their use of basic PV forecasting and rule-based controllers limited the system's ability to anticipate and respond to future fluctuations. The critical importance of accurate PV forecasting has been emphasized by both Ferrarini et al. [13] and Klingler et al. [21], who demonstrated its significant impact on overall system performance.

Despite these advances, significant limitations persist in current research. Most studies employ linear battery models that fail to capture the inherently nonlinear behavior of battery systems. Additionally, insufficient attention has been paid to analyzing PV forecasting and battery states at short time scales. The inadequate consideration of fast dynamics in system operation frequently results in energy imbalances and increased losses during real-time operation.

The proposed framework addresses these limitations by integrating both fast and slow time scales through a comprehensive approach that combines short-term PV forecasting with an electrical circuit-based battery model. This multi-time scale structure enables precise supply-demand balance management while minimizing electricity costs, effectively handling various dynamic elements including PV fluctuations, demand variations, and battery transient responses.

5.3 Proposed Multi-Time Scale Energy Management Framework

This section provides an overview of the proposed multi-time scale energy management framework and discusses the key concepts underlying its multi-time scale structure.

5.3.1 Overview

Figure 5.1 presents a comprehensive overview of the proposed framework, which integrates three principal components: an advanced PV forecasting model, a detailed battery model, and a mathematical formulation of the smart PV system. The framework's primary objective function focuses on electricity cost minimization, with the system utilizing PV generation forecasts and electrical demand profiles as key input parameters. The resulting solution determines the amount of power/energy purchased, battery operational schedules, and temporal allocation of appliance usage.

The framework operates through the following sequential control processes:

1. Generation of PV production profiles utilizing two complementary forecasting approaches: (i) a high-resolution, fine-grained forecasting model for short-term predictions spanning 15–30 minutes with second-level temporal resolution and (ii) a coarse-grained forecasting methodology incorporating meteorological data for extended predictions up to several days.

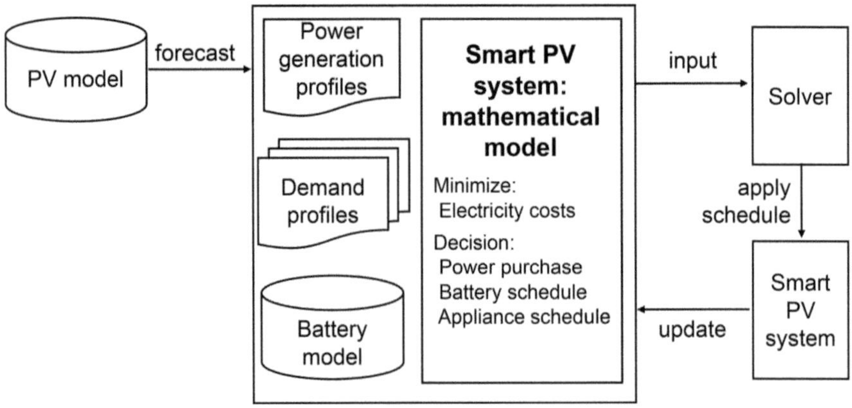

Fig. 5.1 Overview of multi-time scale energy management framework

2. Formulation of a comprehensive mathematical optimization problem that incorporates these forecasted profiles, system constraints, and operational parameters, subsequently solved using specialized mathematical optimization algorithms.
3. Computation of an economically optimal operational schedule that minimizes electricity costs from the utility grid.
4. Evaluation of the derived optimal schedule through numerical simulation, enabling systematic assessment of the control strategy's effectiveness under various operational scenarios and validation of the proposed framework's performance.

These control processes are executed iteratively at each discrete time step in accordance with the MPC methodology, which is elaborated in detail in the subsequent section. This systematic iterative approach facilitates dynamic real-time control, enabling continuous adjustment of the system's energy balance in response to varying operational conditions and external perturbations.

5.3.2 Concept of Model Predictive Control

The proposed framework implements MPC for energy management. MPC has proven effective for controlling complex dynamic systems [15] and has shown particular utility in energy management applications [33]. The core mechanism of MPC involves computing control actions through the minimization of an objective function over a finite prediction horizon, while explicitly considering future system states.

The MPC implementation operates through systematic prediction and optimization steps, as shown in Fig. 5.2. At each sampling time, the controller predicts system behavior over a defined horizon. These predictions are used to formulate

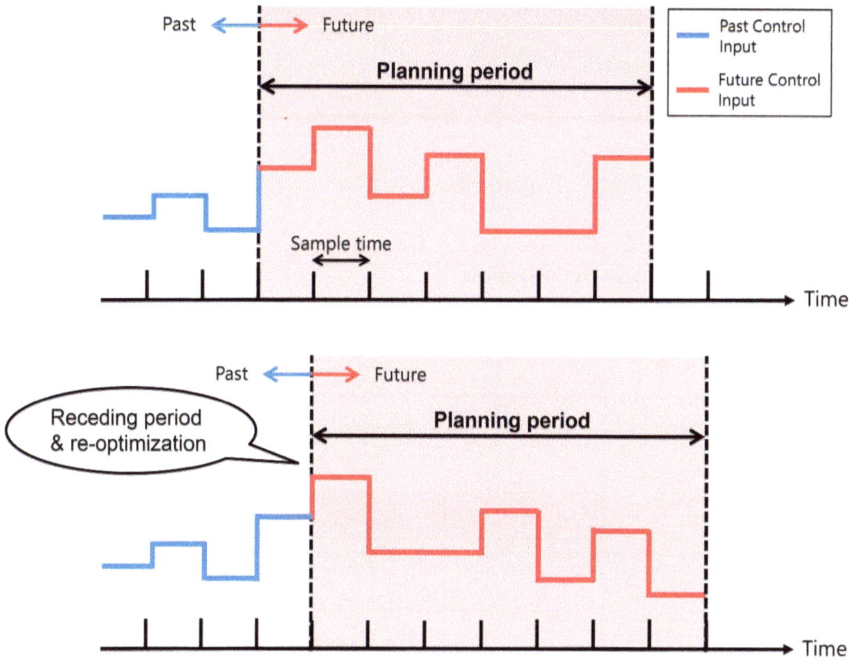

Fig. 5.2 Basic concept of MPC approach

and solve an optimization problem that determines control actions. Following the receding horizon principle, only the first control action is applied, while the optimization is repeated at the next sampling instant with updated measurements and a shifted prediction window.

The feedback structure inherent to MPC provides systematic handling of uncertainties in both system dynamics and external disturbances. This feedback mechanism is particularly relevant for renewable energy systems, where it helps manage variations in load demand and PV generation [31]. The ability to compensate for these uncertainties while maintaining operational constraints makes MPC especially suitable for systems with significant variability in generation and demand.

5.3.3 Multi-Time Scale Approach

The proposed framework adopts a multi-time scale structure to address the inherent temporal disparities in energy management systems. This approach integrates two distinct time scales through a unified optimization framework, as illustrated in Fig. 5.3. At each control point denoted by time index t, the framework executes multi-time scale optimization using the MPC approach, incorporating both forecast information and system states.

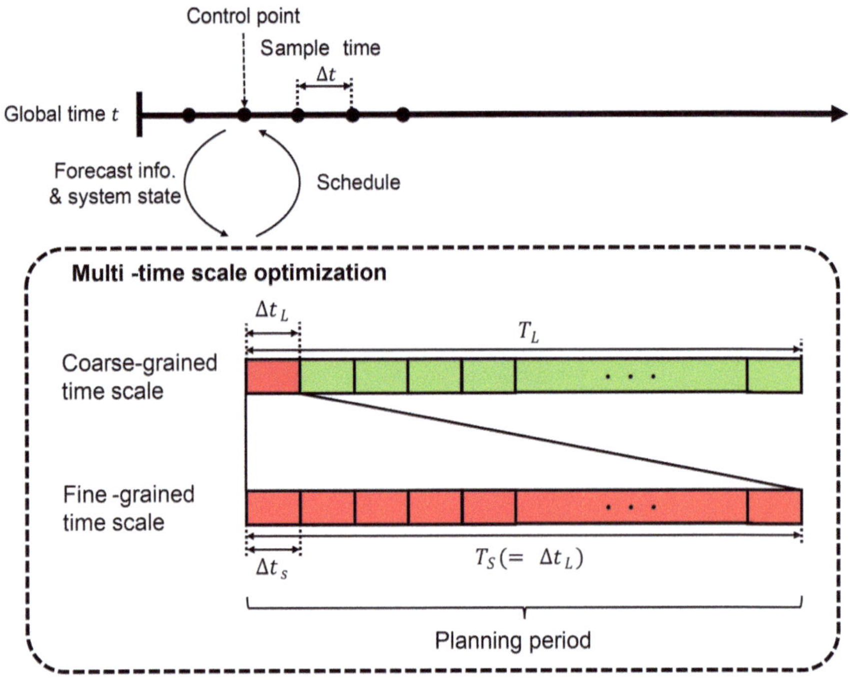

Fig. 5.3 Proposed MPC approach combined with multi-time scale structure

The coarse-grained time scale manages slow dynamics including demand variations and PV generation trends, enabling appliance and battery scheduling over an extended period T_L. While various PV forecasting approaches could potentially be employed, including artificial intelligence-based methods [9], these forecasting models typically demonstrate significantly degraded accuracy at sub-minute resolutions. Consequently, the coarse-grained time resolution Δt_L is set to 15 minutes to maintain forecast reliability while ensuring effective scheduling.

The fine-grained time scale addresses rapid system dynamics, specifically focusing on real-time battery control for energy balancing. This scale utilizes demand schedules from the coarse-grained optimization to compute precise battery operations over a shorter period T_S. The fine-grained time resolution Δt_S must be maintained at the second level to effectively manage rapid system variations. This requirement stems from two fundamental physical constraints: first, PV systems and batteries exhibit electrical time constants of several seconds; second, PV output variations are determined by changes in solar irradiance and PV cell temperature, which respond to environmental factors such as cloud movement and wind speed with time constants typically in the order of several seconds [18, 22].

The multi-time scale approach is essential for three primary reasons: First, it respects the natural time constants of different system components, from rapid PV fluctuations to slower demand variations. Second, it maintains computational tractability by matching optimization time scales to forecast accuracy capabilities.

Third, it enables coordinated control across temporal domains by integrating multiple control objectives into a unified optimization framework, rather than treating them as independent problems.

5.3.4 System Model for Smart PV System

This section describes a detailed explanation of a smart PV system and mathematical formulation.

5.3.4.1 Overview of Smart PV System

In this study, we define a smart PV system as a PV-driven energy community integrating multiple buildings and households. Figure 5.4 illustrates the system architecture, where PV panels and a Li-ion battery system constitute the primary energy components. The battery system provides electrical energy under two conditions: when PV generation is insufficient to meet demand, or when the EMS determines that battery discharge is more economical than energy purchased.

The electrical loads are categorized into two types based on their operational flexibility. Non-shiftable appliances, such as refrigerators and lighting systems, require uninterrupted operation with fixed start times. In contrast, shiftable appliances, including washing machines and dishwashers, allow temporal flexibility in their operation, enabling optimization of their start times within specified intervals.

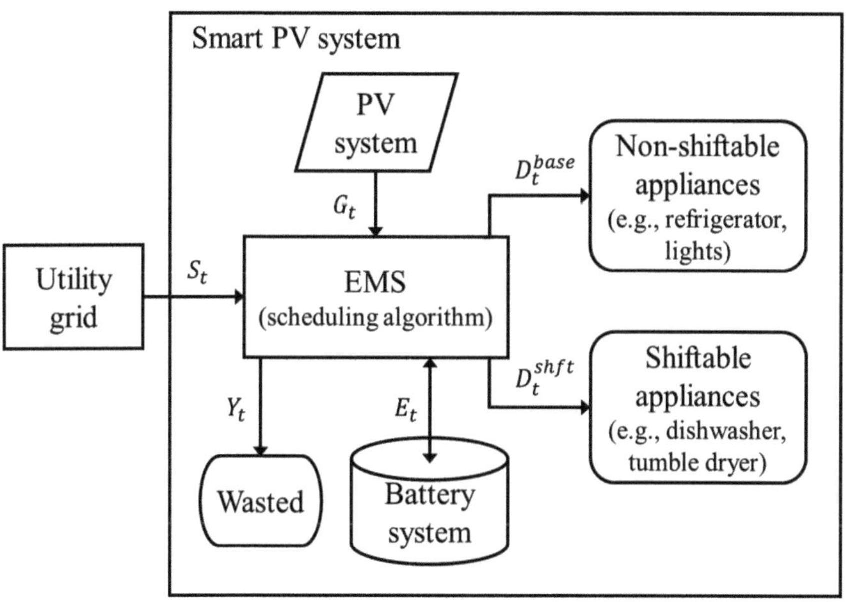

Fig. 5.4 Schematic diagram of smart PV system with mathematical symbols

During periods of energy deficit, the system procures electricity from the utility grid. To maintain grid stability, the system operates under a no-export constraint, prohibiting reverse power flow to the grid. Surplus energy is primarily directed to the battery system and managed loads; any remaining excess is dissipated through dedicated power dissipation components, denoted as "Wasted" in Fig. 5.4.

The EMS implements centralized monitoring and control across all system components through a networked architecture. This configuration enables real-time data acquisition and facilitates information exchange, including meteorological data and component states. The control system optimizes the operation of both shiftable appliances and the battery system to balance PV generation with demand, while adhering to user-defined operational constraints.

The smart PV system can be mathematically formulated through energy balance equations and operational constraints. The fundamental energy conservation equation, which must be satisfied at all time steps t, is expressed as

$$S_t + G_t + E_t = D_t^{base} + D_t^{shft} + Y_t, \ \forall t \tag{5.1}$$

where S_t represents energy purchased from the utility grid, G_t denotes PV energy production, and E_t indicates battery system energy, which is positive for charging and negative for discharging. The terms $D^{base}t$ and $D^{shft}t$ represent the energy consumption of non-shiftable and shiftable appliances, respectively, while Y_t denotes energy wasted within the system.

The system operates under non-negativity constraints for grid purchases S_t and energy dissipation Y_t:

$$0 \leq S_t, \ \forall t \tag{5.2}$$

$$0 \leq Y_t. \ \forall t \tag{5.3}$$

While this formulation captures the essential components for analyzing PV forecasting and battery modeling impacts on EMS performance, the framework maintains extensibility. Additional distributed energy resources (such as wind turbines and electric vehicles) can be incorporated through appropriate mathematical models. Furthermore, the framework can accommodate thermal management objectives, as demonstrated in [45], where similar principles are applied to air conditioning control.

5.3.4.2 Detailed Physics and Neural Network-Based PV Forecasting Model

PV generation exhibits significant fluctuations due to meteorological phenomena, necessitating precise forecasting for effective energy management. This work employs a comprehensive PV nowcasting model, detailed in Chap. 2, which integrates three key components: all-sky camera imagery, physics-based thermal models of PV systems [17], and deep neural network architectures.

The forecasting system operates at two distinct temporal resolutions to address different control requirements. For short-term control, the model generates predictions at 1-second intervals over a 15-minute horizon, with system states updated every minute. This high temporal resolution enables precise battery scheduling while maintaining computational feasibility. Note that, for longer-term planning horizons spanning multiple days, the framework utilizes coarse-grained PV forecasts derived from meteorological data.

5.3.4.3 Shiftable Appliance Model

The operation of shiftable appliances is characterized by four key parameters [35]: (1) operating time; (2) configuration time (T^{conf}), which indicates when the appliance is available for use; (3) deadline (T^{dead}), the latest time by which the appliance must complete its operation; and (4) energy profiles that detail the appliance's energy consumption during operation. The configuration and deadline times are considered user preferences, which means that shiftable appliances must be scheduled to operate within the time window defined by these parameters. Shiftable appliances will start their operation based on the optimized schedule obtained by the proposed framework.

The mathematical formulation captures detailed operational characteristics of each appliance through a multi-phase model, instead of using a single value for appliance power consumption. Each appliance's operating cycle is discretized into P distinct phases, where m and p denote the appliance and phase indices, respectively. The model employs two sets of binary variables: $q_{m,p,t}$ representing operational states (1 if appliance m is in phase p at time t, 0 otherwise), and $r_{m,p,t}$ indicating phase completion status (1 if phase p of appliance m is completed at time t, 0 otherwise).

$$
\begin{aligned}
& D_t^{shft} = \sum_{m=1}^{M} \sum_{p=1}^{P} q_{m,p,t} \cdot D_{m,p}^{app}, \ \forall t \\
& q_{m,p,t} + r_{m,p,t} \leq 1, \ \forall m, p, t \\
& q_{m,p,t-1} - q_{m,p,t} \leq r_{m,p,t}, \ \forall m, p, \ 2 \leq t \leq T \\
& r_{m,p,t-1} \leq r_{m,p,t}, \ \forall m, p, \ 2 \leq t \leq T \\
& q_{m,p,t} \leq r_{m,p-1,t}, \ \forall m, t, \ 2 \leq p \leq P \\
& r_{m,p-1,t} - r_{m,p,t} = q_{m,p,t}, \ \forall m, t, \ 2 \leq p \leq P \\
& \sum_{t=1}^{T} q_{m,p,t} = 1, \ \forall m, p \\
& q_{m,p,t} = 0, \ \forall m, p, \ 1 \leq t \leq T_m^{conf}, \ T_m^{dead} \leq t \leq T
\end{aligned}
\tag{5.4}
$$

where $D_{m,p}^{app}$ represents the energy consumption of appliance m during phase p. This formulation serves two primary functions: it calculates the aggregate energy consumption of shiftable appliances and enforces operational constraints including sequential phase execution and user-defined timing preferences. The optimization problem is solved within the coarse-grained time scale of the framework.

5.3.4.4 Equivalent Circuit-Based Accurate Battery Model

Precise battery modeling is essential for effective energy management system implementation. This work develops a detailed battery module based on an equivalent circuit model [8] that captures nonlinear SOC dynamics and energy losses. This approach addresses limitations of conventional linear models discussed in Sect. 5.2, which oversimplify the relationship between energy loss and charging/discharging processes.

The battery module architecture, shown in Fig. 5.5, consists of identical cells in series-parallel configuration, where N_s and N_p represent the number of series and parallel cells, respectively. The equivalent circuit model has demonstrated high accuracy in representing nonlinear I-V characteristics and temporal behavior [8].

Figure 5.6 presents the detailed equivalent circuit model comprising two main sections. The left section characterizes the battery's state-of-charge (SOC) dynamics through a voltage source V_{SOC} that represents the stored energy level, ranging from 0.0 to 1.0 (0% to 100% charge). The terminal current I_{batt} follows the conventional sign convention: positive for discharge and negative for charge operations. The battery module consists of N_s cells connected in series and N_p parallel strings to achieve the desired voltage and capacity ratings. The nominal capacity C_{nom} scales with the number of parallel strings:

$$C_{nom} = N_p \cdot C_{cell}. \tag{5.5}$$

where C_{cell} represents the capacity of an individual cell. The temporal evolution of SOC is governed by

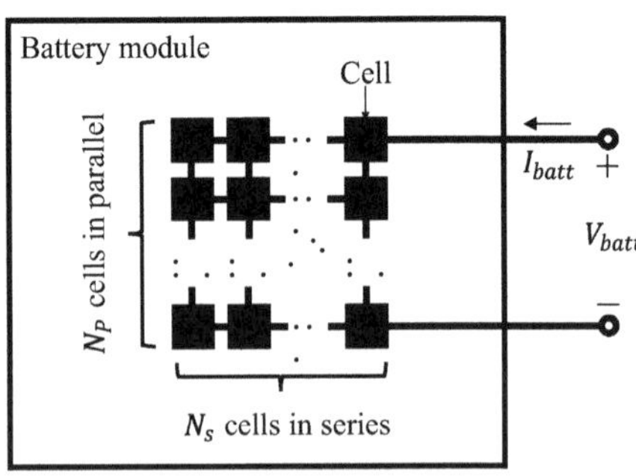

Fig. 5.5 Battery module configuration composed of N_s series cells and N_p parallel cells

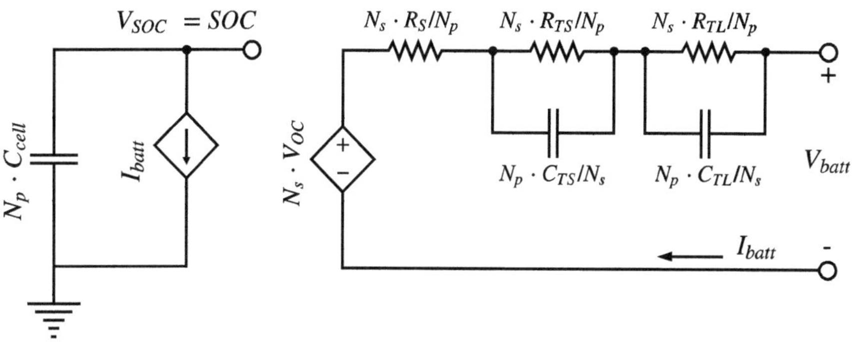

Fig. 5.6 Electrical diagram of equivalent circuit-based battery model composed of $N_s \times N_p$ cells

$$SOC_{t+1} = SOC_t - \int_t^{t+1} \frac{I_{batt}}{C_{nom}} dt. \forall t \tag{5.6}$$

The right section captures the battery's transient response characteristics through two parallel RC networks, which model the dynamic voltage behavior at different time scales. This dual-time-constant approach is essential for accurately representing both rapid and gradual voltage changes during charging and discharging operations. The model incorporates five key SOC-dependent parameters. The open circuit voltage V_{OC} represents the battery's equilibrium potential at a given state of charge, while the series resistance R_S captures the instantaneous voltage drop during current flow. The short-term network (R_{TS}, C_{TS}) models rapid voltage dynamics with time constants typically in the order of seconds, whereas the long-term network (R_{TL}, C_{TL}) represents slower voltage transitions with time constants in the order of minutes.

These parallel RC networks are particularly crucial for modeling the battery's dynamic voltage response to current variations. The short-term network primarily captures immediate voltage recovery effects, while the long-term network accounts for prolonged relaxation phenomena. The SOC-dependent behavior of these parameters is characterized by

$$V_{OC} = a_1 \cdot \exp(a_2 \cdot SOC) + a_3 + a_4 \cdot SOC + a_5 \cdot SOC^2 + a_6 \cdot SOC^3, \tag{5.7}$$

$$R_S = a_7 \cdot \exp(a_8 \cdot SOC) + a_9, \tag{5.8}$$

$$R_{TS} = a_{10} \cdot \exp(a_{11} \cdot SOC) + a_{12}, \tag{5.9}$$

$$R_{TL} = a_{16} \cdot \exp(a_{17} \cdot SOC) + a_{18}, \tag{5.10}$$

$$C_{TS} = a_{13} \cdot \exp(a_{14} \cdot SOC) + a_{15}, \tag{5.11}$$

$$C_{TL} = a_{19} \cdot \exp(a_{20} \cdot SOC) + a_{21}, \tag{5.12}$$

where $\{a_n, \forall n = 1...21\}$ is the set of the battery cell coefficients, which are derived from the literature [8]. Terminal voltage V_{batt} and charging/discharging energy E are given by

$$V_{batt} = N_s \cdot V_{OC} - I_{batt} \cdot \frac{N_s \cdot R_S}{N_p} - U_{TS} - U_{TL}, \tag{5.13}$$

$$E = I_{batt} \cdot V_{batt}/1000, \tag{5.14}$$

where U_{TS} and U_{TL} represent the voltage sources of the left parallel RC branch and the right one, respectively. These values are calculated as follows:

$$\frac{dU_{TS}}{dt} = -\frac{U_{TS}}{R_{TS} \cdot C_{TS}} + I_{batt} \cdot \frac{N_s}{N_p \cdot C_{TS}}, \tag{5.15}$$

$$\frac{dU_{TL}}{dt} = -\frac{U_{TL}}{R_{TL} \cdot C_{TL}} + I_{batt} \cdot \frac{N_s}{N_p \cdot C_{TL}}. \tag{5.16}$$

While this comprehensive model is utilized in system simulation to accurately estimate battery states including SOC transitions and voltage responses, a simplified version is employed in the optimization framework to maintain computational tractability. The complete equivalent circuit model serves as a high-fidelity simulation model for validating system behavior and evaluating control performance, whereas the simplified model, detailed in subsequent sections, provides an efficient representation for real-time optimization while preserving essential battery characteristics.

5.4 Formulation of Proposed Optimization Flow

This section presents the mathematical formulation of the multi-time scale optimization framework, detailing both its structure and solution methodology.

5.4.1 Overview of Control Flow

The proposed framework implements a hierarchical optimization approach to minimize electricity costs in smart PV systems through coordinated scheduling of shiftable appliances, battery operations, and energy purchased. Building on the multi-time scale structure introduced in Sect. 5.3.3, the framework employs MPC across different temporal resolutions.

Figure 5.7 illustrates the framework's control flow, where multiple optimization problems are solved sequentially at intervals of Δt. The optimization hierarchy

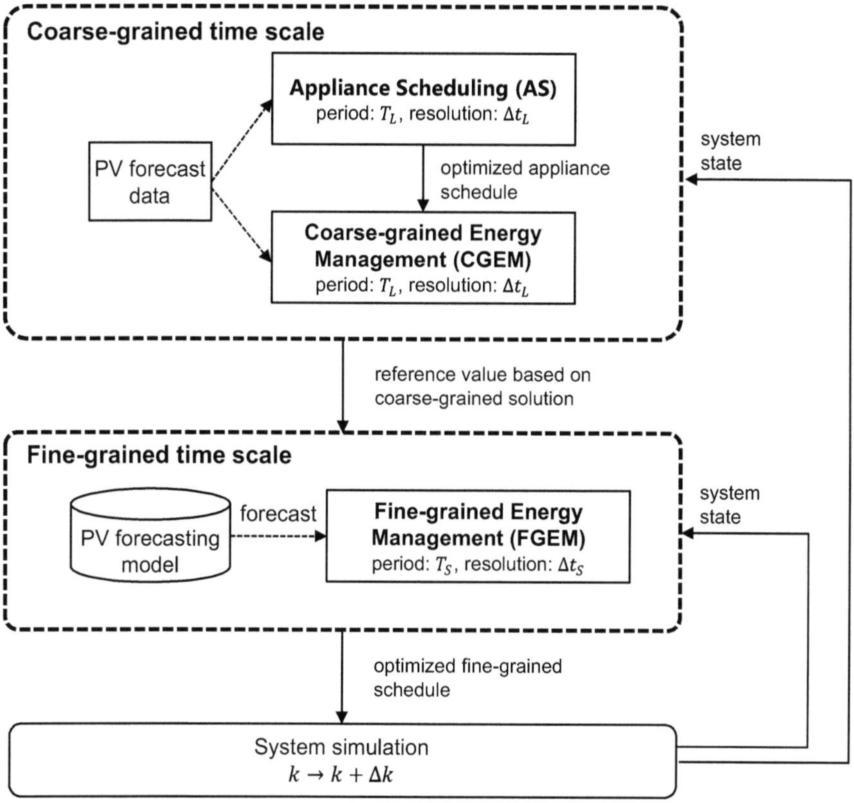

Fig. 5.7 Block diagram of control flow in proposed multi-time scale framework

begins with the coarse-grained time scale, addressing two primary problems: appliance scheduling (AS) and coarse-grained energy management (CGEM). The AS stage determines optimal timing for shiftable loads, while CGEM computes battery system reference trajectories using these appliance schedules as inputs. The CGEM incorporates the equivalent circuit battery model to minimize energy losses. Both stages operate over extended planning horizons to capture daily variations in energy profiles and usage patterns.

The fine-grained time scale subsequently addresses rapid system dynamics through the fine-grained energy management (FGEM) problem. This stage utilizes solutions from the coarse-grained optimization along with short-term PV forecasts detailed in Sect. 5.3.4 to generate precise battery charge/discharge schedules. The resulting control actions are evaluated through detailed system simulation using the complete battery model. Mathematical formulations for each optimization stage are presented in subsequent sections.

5.4.2 Appliance Scheduling

The AS problem operates on the coarse-grained time scale with planning period T_L and resolution Δt_L. The problem is formulated as a MIP optimization to accommodate binary decision variables representing appliance states. The mathematical formulation is expressed as

$$\text{minimize} \quad \sum_{t_L=1}^{T_L} \xi_{t_L} \cdot S_{t_L}, \tag{5.17}$$

$$\text{subject to} \quad (5.1)\text{–}(5.6),\ (5.14),$$

input

$$\{G_{t_L}, D_{t_L}^{base}, \xi_{t_L}\},\ \forall t_L$$

decision variables

$$\{S_{t_L}, Y_{t_L}, I_{batt,t_L}, q_{m,p,t_L}, r_{m,p,t_L}\},\ \forall m,\ p,\ t_L$$

where ξ_{t_L} denotes the time-varying electricity price. To maintain computational tractability while handling integer variables, the formulation employs a simplified battery model that omits nonlinear characteristics of the equivalent circuit. Specifically, the terminal voltage V_{batt} is approximated as a constant nominal value, and the nonlinear equations (5.7)–(5.12), (5.15), and (5.16) describing transient responses and complex I-V characteristics are excluded from the optimization.

The AS solution yields schedules for appliance operation ($D^{shft}{}_{t_L}$, q_{m,p,t_L}), energy purchase (S_{t_L}), battery operation (I_{batt,t_L}), and wasted energy (Y_{t_L}). Only the appliance schedules are retained for implementation, while other variables are recomputed in subsequent optimization stages where the complete battery model is incorporated. This hierarchical approach enables practical computation of appliance schedules while ensuring accurate representation of battery dynamics in later stages.

5.4.3 Coarse-Grained Energy Management (CGEM)

The CGEM constitutes the outer control loop for battery scheduling, operating at the same temporal resolution as the AS stage. This formulation incorporates nonlinear battery characteristics while employing strategic simplifications for computational efficiency. The capacitive elements (C_{TS} and C_{TL}) are excluded from the model as their time constants (20 seconds to 4 minutes) are significantly shorter than the coarse-grained time scale. The resistive components are consolidated into a total resistance R_{total}, yielding

$$R_{total} = R_S + R_{TS} + R_{TL}, \tag{5.18}$$

$$V_{batt} = N_s \cdot V_{OC} - I_{batt} \cdot \frac{N_s \cdot R_{total}}{N_p}. \tag{5.19}$$

The resulting nonlinear programming (NLP) problem, which retains the nonlinear equations (5.7)–(5.10), is formulated as

$$\text{minimize} \quad \sum_{t_L=1}^{T_L} \xi_{t_L} \cdot S_{t_L}, \tag{5.20}$$

subject to (5.1)–(5.3), (5.5)–(5.10), (5.14),(5.18),(5.19),

input

$$\{G_{t_L}, D_{t_L}^{base}, D_{t_L}^{shft}, \xi_{t_L}\}, \ \forall t_L$$

decision variables

$$\{S_{t_L}, Y_{t_L}, I_{batt,t_L}\}, \ \forall t_L$$

While maintaining the same objective of electricity cost minimization, the CGEM generates more accurate schedules than the AS by incorporating battery nonlinearities. The solution encompasses battery operation, purchased energy, and wasted energy schedules. To facilitate fine-grained control, reference energy values E^{ref} are computed from the CGEM solution:

$$E^{ref} = \frac{1}{1000} \cdot I_1^{batt} \cdot V_1^{batt} \cdot \frac{\Delta t_S}{3600}. \tag{5.21}$$

These reference points guide the subsequent FGEM optimization, ensuring balanced battery utilization rather than myopic responses to instantaneous renewable generation.

5.4.4 *Fine-Grained Energy Management (FGEM)*

The FGEM implements the inner control loop for managing rapid PV generation fluctuations through battery control. Operating with a planning period T_S of 15 minutes, which matches the coarse-grained time resolution Δt_L, this stage ensures temporal consistency across the optimization hierarchy.

The FGEM utilizes the appliance schedule $D_{t_S}^{shft}$ from the AS stage while implementing the complete equivalent circuit model to capture fast battery dynamics. To maintain coherence with the coarse-grained solution while allowing for local adjustments, the battery output trajectory is constrained relative to the reference energy E^{ref} by

$$E_{ts} - E^{ref} \leq \varepsilon \cdot |E^{ref}|, \ \forall ts \tag{5.22}$$

where ε represents the acceptable error from E^{ref} and its typical value is 5%.

The FGEM problem results in NLP problem since it includes the nonlinear parts of the battery model. Finally, the mathematical formulation of the FGEM is given by

$$\text{minimize} \quad \sum_{ts=1}^{Ts} \xi_{ts} \cdot S_{ts}, \tag{5.23}$$

$$\text{subject to} \quad (5.1)\text{–}(5.3), (5.5)\text{–}(5.16), (5.22),$$

input

$$\{G_{ts}, D_{ts}^{base}, D_{ts}^{shft}, \xi_{ts}, E^{ref}\}, \ \forall ts$$

decision variables

$$\{S_{ts}, Y_{ts}, I_{batt,ts}\}. \ \forall ts$$

This formulation maintains the objective of electricity cost minimization while determining optimal battery schedules and energy management strategies. The framework evaluates the optimization solutions through detailed system simulation incorporating the complete equivalent circuit model, which captures both slow and fast battery dynamics. Specifically, the simulation implements all nonlinear components including the dual RC networks that model transient responses, SOC-dependent parameters, and voltage characteristics described by Eqs. (5.7)–(5.16). This comprehensive simulation step is essential for two primary reasons: first, internal battery states such as SOC transitions and voltage responses cannot be directly measured in practical systems; second, the interactions between fast battery dynamics and PV fluctuations need to be accurately assessed. The simulation enables quantitative evaluation of the battery's dynamic behavior, including validation of SOC evolution, verification of voltage constraints, and assessment of energy losses under the computed control actions. This detailed validation ensures that the simplified models used in the optimization stages do not compromise the practical feasibility of the derived control strategies.

5.5 Simulation Experiments

This section presents key simulation experiments with practical assumptions to demonstrate the efficacy of the proposed framework. The experimental setup is first described, followed by an exploration of the parameter effects of the proposed method. Additionally, the impact of battery size and PV forecasting error on system performance is investigated. Finally, a performance comparison regarding electricity costs with other baseline methods is conducted.

5.5.1 Simulation Setup

For all simulation experiments, the simulation period is set to 10 days, beginning at midnight. The parameters of the proposed method are outlined below. The coarse-grained time scale covers a 24-hour planning period ($T_L = 96$ [900s]) with a 15-minute resolution ($\Delta t_L = 900$ [s]). In contrast, the fine-grained time scale covers a 15-minute planning period ($T_S = 900$ [s]) with a 1-second resolution ($\Delta t_S = 1$ [s]).

The commercial solver CPLEX v20.1 is used to solve the AS problem, which is a MIP optimization problem. Additionally, the open-source NLP solver IPOPT v3.14 [41] is employed to solve the CGEM and FGEM problems, which are formulated as NLP problems. The simulations are run on a modern laptop equipped with 16 GB RAM and Intel Core-i7 6600U CPU operating at 2.60 GHz.

The parameters of the battery system are listed in Table 5.1. The coefficients $a_n, \forall n = 1...21$ are sourced from the literature [8]. The acceptable error between the battery outputs of the CGEM and FGEM, denoted as ε, is set to 5%. A time-of-use (TOU) pricing scheme commonly used in Japan is applied as input, with a rate of 21.66 ¥/kWh during peak hours (7 a.m.–11 p.m.) and 10.7 ¥/kWh during off-peak hours (11 p.m.–7 a.m.) [20].

PV generation data was measured and collected with a 1-second resolution from June to July 2015 at the University of Oldenburg [2]. For the simulations, data from June 18 to 27 (a 10-day period) is selected as input, as illustrated in Fig. 5.8. To simulate a 15 kWp PV system, the PV generation profiles are amplified by a constant scaling factor. It is important to note that the weather during these simulated

Table 5.1 Parameter settings of battery system

Description	Symbol	Value
Initial SOC	SOC^{init}	0.5 (50%)
Terminal SOC	SOC^{term}	0.5 (50%)
Min. SOC	$\underline{SOC}$	0.2 (20%)
Max. SOC	$\overline{SOC}$	1 (100%)
Min. current	$\underline{I_{batt}}$	$-0.5 \cdot C_{nom}$ (50% of capacity)
Max. current	$\overline{I_{batt}}$	$0.5 \cdot C_{nom}$ (50% of capacity)
Number of series cells	N_s	25
Number of parallel cells	N_p	191
Nominal voltage	V_{cell}	4.1 [V]
Nominal capacity	C_{cell}	0.85 [Ah]
Battery capacity	–	15 [kWh]

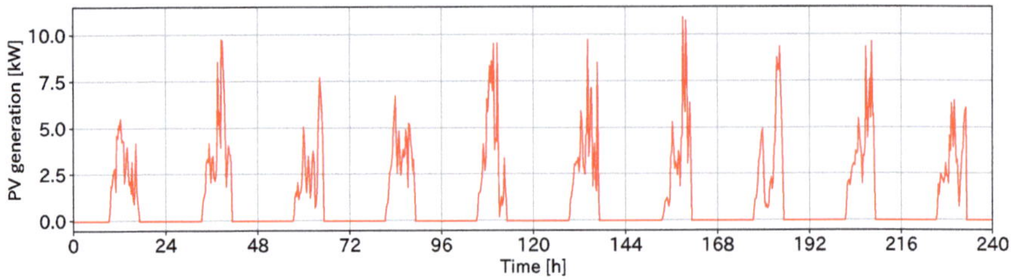

Fig. 5.8 PV generation used in simulation, measured at the University of Oldenburg from June 18 to July 27

Table 5.2 Parameters for smart appliances, washing machine, tumble dryer, and dishwasher

Appliance	Total energy	Operating time	Configuration time	Shiftable time
Washing machine	0.22 kWh	45 min	8 a.m.–10 a.m.	7 h
Tumble dryer	1.86 kWh	75 min	8 a.m.–10 a.m.	7 h
Dishwasher	1.88 kWh	75 min	12 p.m.–15 p.m.	8 h

days was predominantly cloudy, which tends to cause significant fluctuations in PV generation and makes accurate PV forecasting more challenging. Due to these cloudy conditions, high forecasting errors are expected in this simulation, representing a worst-case scenario and demonstrating the worst-case performance of the proposed method. For PV forecasting, the PV nowcasting model [2] is utilized to obtain fine-grained PV generation forecasts, which show an average forecasting error of 12%, even under cloudy conditions. Since this work does not implement a specific forecasting model for the coarse-grained time scale, the coarse-grained PV forecast is manually generated by adding Gaussian-distributed errors to the measured profiles, resulting in an average error of 20%.

The demand profiles of non-shiftable appliances are derived from the Dutch Residential Energy Dataset (DRED) [39], which was collected with a 1-second resolution from July to December 2015. For the simulation, demand profiles spanning 10 days are extracted from DRED, specifically from July 5 to July 14. These profiles are scaled up by a constant factor, resulting in an average daily consumption of non-shiftable appliances set to 50.1 kWh. The parameters for shiftable appliances are detailed in Table 5.2, which includes three types of shiftable appliances, dishwashers, washing machines, and tumble dryers, each operated once per day. There are 4 appliances of each type, totaling 12 shiftable appliances. The configuration times for these appliances are randomly generated within the ranges specified in Table 5.2. The deadlines for appliance operation are determined by adding the shiftable time to the configuration time. The dataset [34] provides power profiles for shiftable appliances with a 1-second resolution. Representative examples of the power profiles for a washing machine, tumble dryer, and dishwasher are shown in Figs. 5.9 and 5.10, respectively.

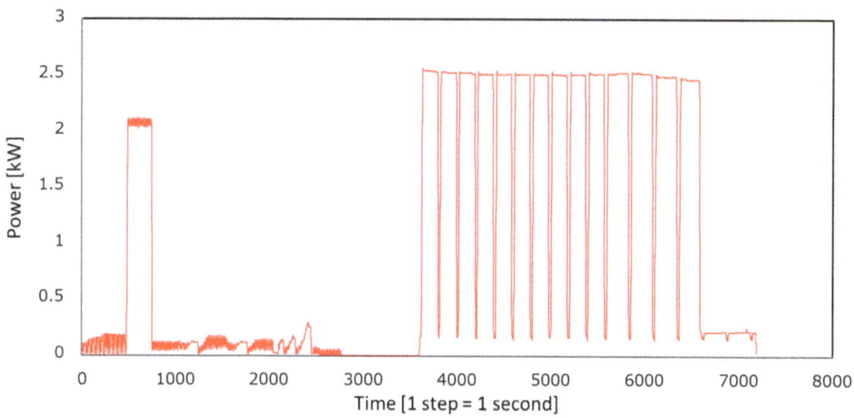

Fig. 5.9 Power profiles of washing machine (– 3000 s) and tumble dryer (3000–7500 s)

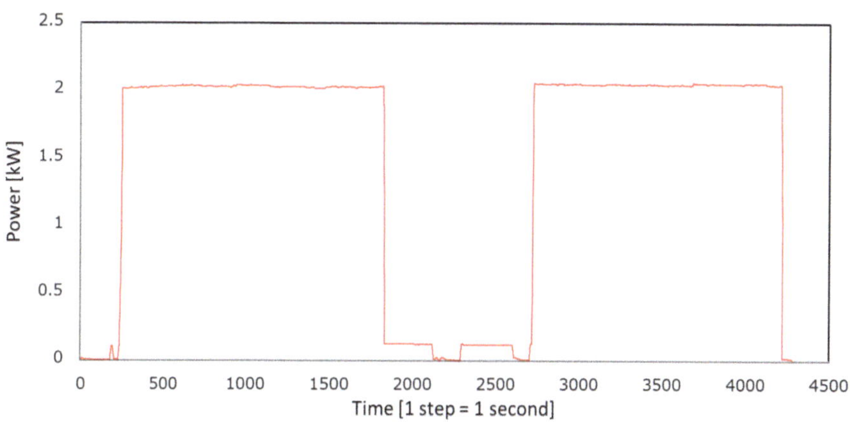

Fig. 5.10 Power profiles of dishwasher

5.5.2 Results

5.5.2.1 Comparison with Baseline Scheduling Methods

To evaluate the efficacy of appliance and battery scheduling, this section compares the proposed framework to several baseline methods, described as follows:

1. **Using Shiftable Appliances As Soon As Possible (ASAP):** In this method, shiftable appliances are not scheduled through optimization. Instead, appliances are immediately turned on as soon as their configuration time arrives. The battery system schedules are optimized using the CGEM and FGEM.

Table 5.3 Electricity cost and its improving rates of proposed method within 10 days compared to three methods, ASAP, NBS, and ASAP-NBS

Method	Proposed	ASAP	NBS	ASAP-NBS
Electricity cost within 10 days [¥]	3004	3319	4408	5784
Improving rate of electricity cost of proposed	–	9.5%	31.9%	48.1%

Table 5.4 Electricity cost and computational time for different lengths of planning period

Planning period [h]		Computational time [s]		
T_L	Electricity costs [¥]	AS	CGEM	FGEM
6	3194	0.95	0.19	3.11
12	3106	1.63	0.27	3.03
24	3004	4.32	0.43	3.51
36	3044	8.65	0.63	3.12
48	3055	17.59	0.94	3.24

2. **No Battery Scheduling (NBS):** In this approach, the CGEM and FGEM are excluded from the proposed method. Only the AS stage is employed, and the battery system operates on a fixed schedule, charging at a constant C-rate of 10% from 11 p.m. to 7 a.m. and discharging at a constant C-rate of 5% from 7 a.m. to 11 p.m.
3. **ASAP-NBS:** This method combines the principles of ASAP and NBS, resulting in no optimization problem being solved.

Table 5.3 presents the results of electricity costs over a 10-day period, along with the improvement rates of the proposed framework compared to other methods. The proposed framework achieved the lowest electricity cost among all methods, with a maximum improvement rate of 48.1%. Appliance scheduling effectively helps to bridge the energy gap between generation and demand, leading to reduced electricity costs. Battery scheduling also plays a crucial role in minimizing electricity costs. When the battery system is charged and discharged at a constant current, as in the NBS method, it cannot effectively balance renewable generation and demand, resulting in increased energy purchases to meet energy requirements.

5.5.2.2 Effect of Planning Period on Coarse-Grained Time Scale

In this section, the effect of the planning period on the coarse-grained time scale is investigated. For the proposed method, the planning period T_L is varied from 6 hours to 48 hours.

Table 5.4 shows the electricity costs over a 10-day period and the average computational time for each optimization problem. As seen in the table, electricity costs decrease as the planning period increases up to 24 hours but begin to rise

Table 5.5 Average AS's computational time for different numbers of smart appliances

Shiftable appliances per day	Computational times of AS [s]
6	2.13
12	4.32
18	6.67
24	8.79
30	10.39

when the planning period exceeds 24 hours. This trend is due to changes in the battery schedule solutions for the 36-hour and 48-hour planning periods. The SOC of the battery at the end of the 10-day simulation is 53.1% when using a 24-hour planning period. However, the SOC increases to 65.1% and 65.8% when using 36-hour and 48-hour planning periods, respectively. This indicates that with longer planning periods, more energy remains stored in the battery. Additionally, when a 36-hour or 48-hour planning period is used, some shiftable appliances on the last day are scheduled outside the evaluation period. In contrast, using a 24-hour planning period allows these appliances to be scheduled within the same day.

Furthermore, computational time increases as the planning period becomes longer. For the 48-hour planning period, the computational time for the AS problem increases significantly due to the need to schedule more smart appliances. However, the total computational time remains much less than the length of the time resolution $\Delta t_L = 900$ [s], indicating that the proposed framework is applicable for all simulated planning periods. In this case study, a 24-hour planning period is the most efficient, achieving a good balance between computational complexity and solution quality.

5.5.2.3 Effect of Number of Smart Appliances

Next, the impact of the number of smart appliances on computational time was examined. The number of each type of smart appliance was increased from 2 to 10, resulting in a total of 6 to 30 appliances. Table 5.5 presents the average computational time for each optimization problem, which increases as more appliances are added. However, the computational time remains sufficiently short, consistently meeting the requirements of $\Delta t_L = 900$ [s]. Therefore, the proposed framework is applicable to smart PV systems comprising multiple buildings with up to 30 or more smart appliances.

5.5.2.4 Effect of PV Forecasting Error and Battery Size

In this section, the system performance was analyzed under different battery sizes and PV forecasting errors. Various coarse-grained forecasts of PV generation were

Table 5.6 Electricity cost values in ¥with different PV forecasting errors and battery capacities

Battery capacity [kWh]	Perfect forecasting	Forecasting scheme for coarse-grained					
		Energy forecasting (error <12%)			Power forecasting (error <20%)		
		Forecast error for fine-grained			Forecast error for fine-grained		
		20%	30%	40%	20%	30%	40%
3	3450	3558	3529	3579	3610	3570	3612
6	3146	3309	3292	3360	3437	3404	3477
9	3025	3181	3227	3272	3334	3381	3420
12	2821	3037	3118	3171	3202	3289	3353
15	2739	3004	3094	3185	3182	3271	3379
18	2719	3027	3092	3205	3193	3281	3384

employed, with average forecasting errors of 20%, 30%, and 40%. For the fine-grained time scale, two different forecasting schemes were compared, energy forecasting [2] and power forecasting, both of which were also presented in [2]. The average forecasting error for the 15-minute periods was 12% for energy forecasting and 20% for power forecasting. Additionally, to assess the impact of perfect forecasting, a scenario with a forecasting error of 0% was included, assuming the PV forecast method works ideally. The battery capacity was varied from 3 kWh to 18 kWh to investigate the effect of battery size on system performance.

Figure 5.11 shows the electricity costs over a 10-day period with varying PV forecasting errors and battery capacities. The corresponding electricity cost values are provided in Table 5.6. In Fig. 5.11, the black line represents the electricity costs under perfect forecasting conditions. The blue and red lines indicate the results of power forecasting and energy forecasting, respectively. As illustrated by these results, electricity costs do not significantly improve when the battery capacity exceeds 12 kWh. This is because errors in predicting future PV generation profiles have a substantial impact on performance as battery size increases. In other words, incorrect battery operation due to PV forecasting errors offsets the benefits of reduced electricity costs achieved by a larger battery capacity. From the perspective of high initial costs, opting for a larger battery size may not be advisable.

The forecasting error on the coarse-grained time scale has a significant impact on electricity costs. Specifically, smaller forecasting errors result in lower electricity costs. The impact of coarse-grained forecasting errors on electricity costs is minimal when the battery capacity is 3 or 6 kWh. However, when the battery capacity exceeds 9 kWh, the difference in electricity costs between varying forecasting errors becomes more pronounced. Furthermore, the effect of fine-grained PV forecasting errors on electricity costs is greater than that of coarse-grained errors. A 10% improvement in fine-grained forecasting accuracy can lead to a 30–50% reduction in required battery size to achieve similar electricity cost savings. Therefore, the accuracy of the forecasting scheme for the fine-grained time scale is a crucial factor in the performance of the EMS.

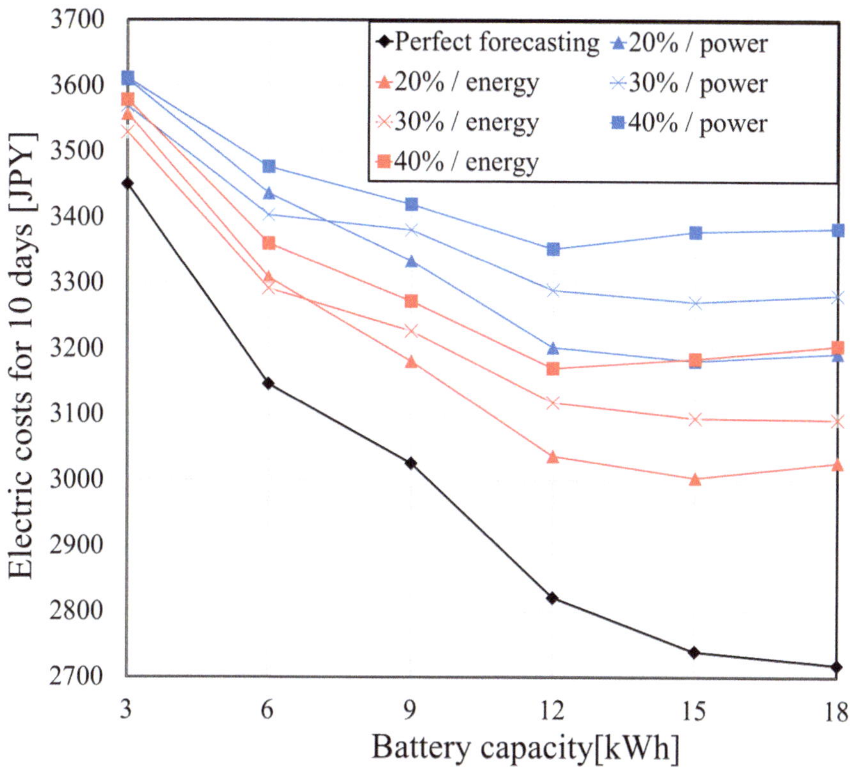

Fig. 5.11 Electricity costs within 10 days for different scenarios of PV forecasting errors and different battery capacities, where the value is the ratio between forecasting error in coarse-grained and scheme in fine-grained

5.5.2.5 Effect of Multi-Time Scale Structure

To determine the necessity of the multi-time scale structure, the proposed framework was compared to two single-time scale methods:

1. **Only Coarse-grained Optimization (OC):** In this method, only the AS and CGEM are executed on the coarse-grained time scale. The solution is directly applied to the system without solving the FGEM. OC utilizes a 24-hour planning period with a 15-minute resolution.
2. **Only Fine-grained Optimization (OF):** This approach only executes the FGEM on the fine-grained time scale. Smart appliances are operated as soon as possible, similar to the ASAP method, and the CGEM is not used. OF operates with a 15-minute planning period and a 1-second resolution.

Note that the parameters of OC and OF, namely, the planning period and time resolution, are chosen based on the characteristics of the available PV forecasting methods. As mentioned in Sect. 5.3.3, the accuracy of long-term PV forecasting

Table 5.7 Comparison of electricity cost and its improving rate for proposed method, OC, and OF

Method	Proposed	OC	OF
Electricity cost for 10 days [¥]	3004	3183	5717
Improving rate of electricity costs of Proposed	–	5.6%	47.5%

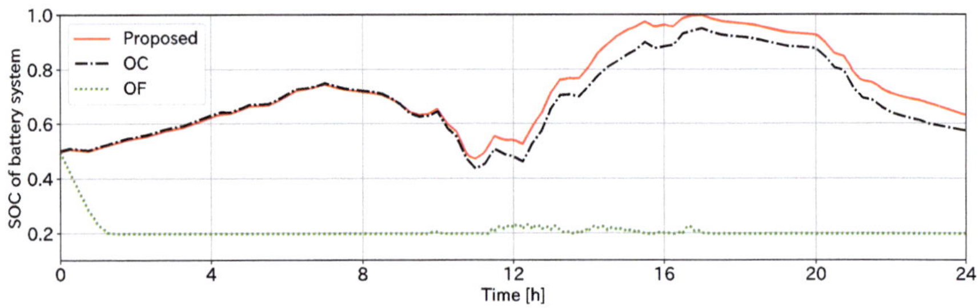

Fig. 5.12 SOC profiles of June 18 for three methods, Proposed, OC, and OF, with 15 kWh battery system

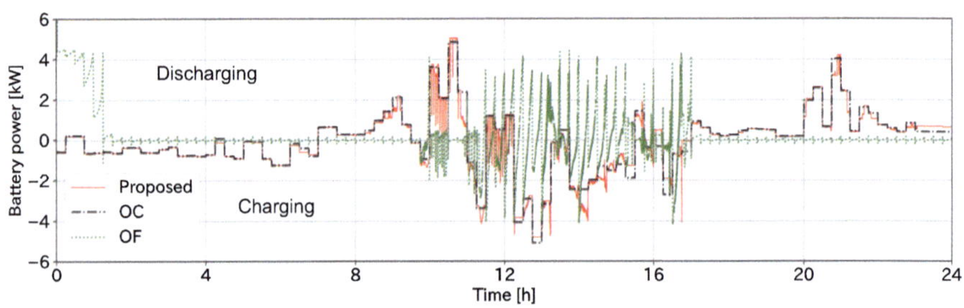

Fig. 5.13 Battery power profiles of June 18 for three methods, Proposed, OC, and OF, with 15 kWh battery system; Positive value represents discharging, otherwise charging

decreases significantly when the time resolution is less than a few minutes (e.g., less than 5 minutes) [9]. Conversely, short-term PV forecasting, which is based on sky images and physics-based models, can provide accurate profiles with a 1-second resolution; however, its accuracy degrades rapidly when the forecasting period extends beyond 15 minutes [2]. To align the target time scales with each PV forecasting approach and perform the comparative experiment under realistic assumptions, the parameters of OC and OF are set as described above.

Table 5.7 presents the results of electricity costs over a ten-day period and the improvement rate of the proposed method compared to the OC and OF methods. The proposed framework achieved the best performance, with a maximum improvement rate of 47.5% in electricity costs. This result demonstrates the effectiveness of the multi-time scale structure in reducing electricity costs.

Figures 5.12 and 5.13 show the SOC profiles and battery power profiles for the same day. The results for the OF method clearly demonstrate myopic optimization, where the battery system is discharged to reduce electricity costs as long as there is energy remaining in the battery. Since OF only considers the upcoming 15 minutes (T_S) without incorporating the coarse-grained time scale, its solution is not optimized for long-term changes in PV generation and demand load. In comparison, the proposed method achieves a higher SOC level than both the proposed method and OC. While OC provides a solution that accounts for long-term changes in energy profiles, accumulated errors due to PV fluctuations on the fine-grained time scale result in fewer opportunities to charge the battery using PV generation. On the other hand, the battery solution of the proposed framework considers long-term changes by adhering to reference values from the coarse-grained time scale and compensates for rapid PV fluctuations by solving FGEM with fine-grained PV forecasting data.

Overall, the multi-time scale structure leads to superior performance in energy management. The proposed framework primarily focuses on managing the fast and slow dynamics of PV generation and the battery system. Although energy demand is also highly volatile, its fluctuations occur over many minutes, which is distinct from the time scales of PV generation and the battery system. Consequently, fluctuations in energy demand can be effectively managed within the coarse-grained optimization loop of the multi-time scale approach. As indicated by the simulation results, the quick response of battery charging and discharging allows the system to efficiently handle energy demand fluctuations.

5.6 Summary

This chapter presented a multi-time scale energy management framework for a smart PV system. An MPC strategy, using PV generation forecasting as input, was adopted to address the high fluctuations in PV generation. The proposed framework employs a multi-time scale structure, comprising both coarse-grained and fine-grained time scales. For the coarse-grained time scale, the framework optimizes the schedules of smart appliances to adjust their operating times and optimizes battery outputs to manage daily changes in the energy balance. For the fine-grained time scale, the battery system schedule is fine-tuned by incorporating an accurate battery model and a fine-grained PV forecasting model. The proposed framework effectively solves three internally connected optimization problems, considering both fast and slow system dynamics. Simulation results indicate that the proposed framework can reduce electricity costs by up to 47.5% compared to baseline methods across various scenarios. Additionally, the impact of PV forecasting error and battery capacity on the framework's performance was analyzed. The results suggest that integrating an accurate PV forecasting model could further reduce electricity costs, even with a smaller battery system. In summary, the proposed framework provides the EMS with real-time control capabilities, resulting in a highly accurate solution and enabling a reduction in the size of an expensive battery system.

Future work should address several key limitations of the current framework. A primary area for enhancement is the incorporation of workload-dependent battery dynamics. The current framework does not account for critical battery characteristics such as degradation patterns and complex SOC variations. These battery dynamics typically exhibit time constants at the minute level, substantially longer than the 1-second resolution employed in the fine-grained time scale. To address this limitation, an intermediate time scale – a semi-fine-grained level operating at minute-level resolution with planning periods exceeding 1 hour – will be considered.

References

1. Abreu, A., Bourdais, R., Guéguen, H.: Hierarchical model predictive control for building energy management of hybrid systems. IFAC-PapersOnLine **51**(16), 235–240 (2018)
2. Anagnostos, D., Schmidt, T., Cavadias, S., Soudris, D., Poortmans, J., Catthoor, F.: A method for detailed, short-term energy yield forecasting of photovoltaic installations. Renew. Energy **130**, 122–129 (2019)
3. Anees, A., Chen, Y.P.P.: True real time pricing and combined power scheduling of electric appliances in residential energy management system. Appl. Energy **165**, 592–600 (2016)
4. Arcos-Aviles, D., Pascual, J., Marroyo, L., Sanchis, P., Guinjoan, F.: Fuzzy logic-based energy management system design for residential grid-connected microgrids. IEEE Trans. Smart Grid **9**(2), 530–543 (2018)
5. Azuatalam, D., Paridari, K., Ma, Y., Förstl, M., Chapman, A.C., Verbič, G.: Energy management of small-scale PV-battery systems: a systematic review considering practical implementation, computational requirements, quality of input data and battery degradation. Renew. Sustain. Energy Rev. **112**, 555–570 (2019)
6. Carli, R., Cavone, G., Ben Othman, S., Dotoli, M.: IoT based architecture for model predictive control of HVAC systems in smart buildings. Sensors **20**(3) (2020)
7. Chen, C., Duan, S., Cai, T., Liu, B., Hu, G.: Smart energy management system for optimal microgrid economic operation. IET Renew. Power Generation **5**(3), 258–267 (2011)
8. Chen, M., Rincon-Mora, G.A.: Accurate electrical battery model capable of predicting runtime and I-V performance. IEEE Trans. Energy Conversion **21**(2), 504–511 (2006)
9. Diagne, M., David, M., Lauret, P., Boland, J., Schmutz, N.: Review of solar irradiance forecasting methods and a proposition for small-scale insular grids. Renew. Sustain. Energy Rev. **27**, 65–76 (2013)
10. Dorahaki, S., Dashti, R., Shaker, H.R.: Optimal energy management in the smart microgrid considering the electrical energy storage system and the demand-side energy efficiency program. J. Energy Storage **28**, 101229 (2020)
11. Elkazaz, M., Sumner, M., Naghiyev, E., Pholboon, S., Davies, R., Thomas, D.: A hierarchical two-stage energy management for a home microgrid using model predictive and real-time controllers. Appl. Energy **269**, 115118 (2020)
12. Erdinc, O.: Economic impacts of small-scale own generating and storage units, and electric vehicles under different demand response strategies for smart households. Appl. Energy **126**, 142–150 (2014)
13. Ferrarini, L., Mantovani, G., Costanzo, G.T.: A distributed model predictive control approach for the integration of flexible loads, storage and renewables. In: Proceedings of the 2014 IEEE 23rd International Symposium on Industrial Electronics (ISIE), pp. 1700–1705. IEEE, Piscataway (2014)
14. Gan, L.K., Zhang, P., Lee, J., Osborne, M.A., Howey, D.A.: Data-driven energy management system with gaussian process forecasting and MPC for interconnected microgrids. IEEE Trans. Sustain. Energy **12**(1), 695–704 (2021)

15. García, C.E., Prett, D.M., Morari, M.: Model predictive control: theory and practice—a survey. Automatica: J. IFAC Int. Fed. Autom. Control **25**(3), 335–348 (1989)
16. Godina, R., Rodrigues, E.M.G., Pouresmaeil, E., Matias, J.C.O., Catalão, J.P.S.: Model predictive control home energy management and optimization strategy with demand response. NATO Adv. Sci. Inst. Ser. E: Appl. Sci. **8**(3), 408 (2018)
17. Goverde, H., Anagnostos, D., Govaerts, J., Manganiello, P., Voroshazi, E., Szlufcik, J., Catthoor, F., Poortmans, J., Baert, K., Driesen, J.: Accurately simulating PV energy production: exploring the impact of module build up. In: Proceedings of the 33rd European Photovoltaics Science and Engineering Conference (EUPVSEC), pp. 1643–1646 (2017)
18. Goverde, H., Anagnostos, D., Herteleer, B., Govaerts, J., Baert, K., Aldalali, B., Catthoor, F., Driesen, J., Poortmans, J.: Model requirements for accurate short term energy yield predictions during fast-varying weather conditions. In: Proceedings of the 31st European Photovoltaic Solar Energy Conference and Exhibition (EUPVSEC), pp. 1556–1559 (2015)
19. Jin, X., Wu, J., Mu, Y., Wang, M., Xu, X., Jia, H.: Hierarchical microgrid energy management in an office building. Appl. Energy **208**, 480–494 (2017)
20. Kansai Electric Power Co., Inc.: The Time-of-Use electricity price in the Kansai electric power company in Japan. https://kepco.jp/ryokin/menu/jikanbetsu/. Accessed 1 Dec 2022
21. Klingler, A.L., Teichtmann, L.: Impacts of a forecast-based operation strategy for grid-connected PV storage systems on profitability and the energy system. Solar Energy **158**, 861–868 (2017)
22. Lappalainen, K., Valkealahti, S.: Analysis of shading periods caused by moving clouds. Solar Energy **135**, 188–196 (2016)
23. Lefort, A., Bourdais, R., Ansanay-Alex, G., Guéguen, H.: Hierarchical control method applied to energy management of a residential house. Energy Build. **64**, 53–61 (2013)
24. Liang, X.: Emerging power quality challenges due to integration of renewable energy sources. IEEE Trans. Ind. Appl. **53**(2), 855–866 (2017)
25. Lokeshgupta, B., Sivasubramani, S.: Multi-objective home energy management with battery energy storage systems. Sustain. Cities Soc. **47**, 101458 (2019)
26. Lu, Q., Zhang, Z., Lü, S.: Home energy management in smart households: optimal appliance scheduling model with photovoltaic energy storage system. Energy Rep. **6**, 2450–2462 (2020)
27. Lund, H., Østergaard, P.A., Connolly, D., Mathiesen, B.V.: Smart energy and smart energy systems. Energy **137**, 556–565 (2017)
28. Mariano-Hernández, D., Hernández-Callejo, L., Zorita-Lamadrid, A., Duque-Pérez, O., Santos García, F.: A review of strategies for building energy management system: model predictive control, demand side management, optimization, and fault detect & diagnosis. J. Build. Eng. **33**, 101692 (2021)
29. Merabet, A., Tawfique Ahmed, K., Ibrahim, H., Beguenane, R., Ghias, A.M.Y.M.: Energy management and control system for laboratory scale microgrid based Wind-PV-Battery. IEEE Trans. Sustain. Energy **8**(1), 145–154 (2017)
30. Olatomiwa, L., Mekhilef, S., Ismail, M.S., Moghavvemi, M.: Energy management strategies in hybrid renewable energy systems: a review. Renew. Sustain. Energy Rev. **62**, 821–835 (2016)
31. Parisio, A., Rikos, E., Glielmo, L.: A model predictive control approach to microgrid operation optimization. IEEE Trans. Control Syst. Technol. **22**(5), 1813–1827 (2014)
32. Parisio, A., Wiezorek, C., Kyntäjä, T., Elo, J., Strunz, K., Johansson, K.H.: Cooperative MPC-based energy management for networked microgrids. IEEE Trans. Smart Grid **8**(6), 3066–3074 (2017)
33. Ramos Ruiz, G., Lucas Segarra, E., Fernández Bandera, C.: Model predictive control optimization via genetic algorithm using a detailed building energy model. Energies **12**(1), 34 (2018)
34. Reinhardt, A., Baumann, P., Burgstahler, D., Hollick, M., Chonov, H., Werner, M., Steinmetz, R.: On the accuracy of appliance identification based on distributed load metering data. In: Proceedings of the 2012 Sustainable Internet and ICT for Sustainability (SustainIT), pp. 1–9 (2012)

35. Sadeghianpourhamami, N., Demeester, T., Benoit, D.F., Strobbe, M., Develder, C.: Modeling and analysis of residential flexibility: timing of white good usage. Appl. Energy **179**, 790–805 (2016)
36. Sou, K.C., Weimer, J., Sandberg, H., Johansson, K.H.: Scheduling smart home appliances using mixed integer linear programming. In: Proceedings of the 2011 50th IEEE Conference on Decision and Control and European Control Conference (ECC), pp. 5144–5149 (2011)
37. Szumanowski, A., Chang, Y.: Battery management system based on battery nonlinear dynamics modeling. IEEE Trans. Veh. Technol. **57**(3), 1425–1432 (2008)
38. Terlouw, T., AlSkaif, T., Bauer, C., van Sark, W.: Optimal energy management in all-electric residential energy systems with heat and electricity storage. Appl. Energy **254**, 113580 (2019)
39. Uttama Nambi, A.S.N., Reyes Lua, A., Prasad, V.R.: LocED: location-aware energy disaggregation framework. In: Proceedings of the 2nd ACM International Conference on Embedded Systems for Energy-Efficient Built Environments (BuildSys), pp. 45–54 (2015)
40. Varun, Prakash, R., Bhat, I.K.: Energy, economics and environmental impacts of renewable energy systems. Renew. Sustain. Energy Rev. **13**(9), 2716–2721 (2009)
41. Wächter, A., Biegler, L.T.: On the implementation of an interior-point filter line-search algorithm for large-scale nonlinear programming. Math. Program. Publ. Math. Program. Soc. **106**(1), 25–57 (2006)
42. Watari, D., Taniguchi, I., Goverde, H., Manganiello, P., Shirazi, E., Catthoor, F., Onoye, T.: Multi-time scale energy management framework for smart PV systems mixing fast and slow dynamics. Appl. Energy **289**, 116671 (2021)
43. Zhang, Y., Lundblad, A., Campana, P.E., Benavente, F., Yan, J.: Battery sizing and rule-based operation of grid-connected photovoltaic-battery system: a case study in Sweden. Energy Conver. Manag. **133**, 249–263 (2017)
44. Zhang, Y., Wang, R., Zhang, T., Liu, Y., Guo, B.: Model predictive control-based operation management for a residential microgrid with considering forecast uncertainties and demand response strategies. IET Gener. Transmission Distrib. **10**(10), 2367–2378 (2016)
45. Zhao, D., Watari, D., Ozawa, Y., Taniguchi, I., Suzuki, T., Shimoda, Y., Onoye, T.: Data-driven online energy management framework for HVAC systems: an experimental study. Appl. Energy **352**, 121921 (2023)

Correction to: Energy Production, Load and Battery Management Framework with Supporting Methods for Smart Microgrids

Francky Catthoor, Dafang Zhao, Ittetsu Taniguchi, Pavlos S. Georgilakis, Dimitrios Soudris, Kostas Siozios, and Andreas Kazantzidis

Correction to:
F. Catthoor et al. (eds.), *Energy Production, Load and Battery Management Framework with Supporting Methods for Smart Microgrids*, https://doi.org/10.1007/978-3-031-92025-7

The book was inadvertently published with incorrect author order.

For this book, the order of the volume editors needs to be revised:

The order of the volume editors was requested to be changed, but the change was not carried out in proof. The correct order should be: Catthoor, Zhao, Taniguchi, Georgilakis, Soudris, Siozios, and Kazantzidis.

The updated version of this book can be found at
https://doi.org/10.1007/978-3-031-92025-7

F. Catthoor et al. (eds.), *Energy Production, Load and Battery Management Framework with Supporting Methods for Smart Microgrids*,
https://doi.org/10.1007/978-3-031-92025-7_6

Glossary

$E_k^{(t)}$	Objective function value at state k during step t of SAMURAI
$E_{ref}^{(t)}$	Reference objective function value at step t of SAMURAI
$F(\cdot)$	Cumulative distribution function used for CRPS calculation
$M^{(t)}$	Total number of explored states at step t of SAMURAI
$N_k^{(t)}$	Number of states explored up to state k at step t of SAMURAI
$N_{k,uphill}^{(t)}$	Number of uphill states explored up to state k at step t of SAMURAI
$T^{(t)}$	Temperature value at step t during annealing with SAMURAI
c_i	Binary parameter indicating whether target value i lies inside the corresponding PI
c_1, c_2	Acceleration constants for velocity update during optimization with PSO
cl	Confidence level of PIs
g_{best}	Best position explored in swarm during optimization with PSO
h	Constant used for CWC calculation
$l(\cdot)$	Heaviside function used for CRPS calculation
l_i	Lower bound of PI generated for sample i
$p_{best,n}$	Best position explored by particle n during optimization with PSO
u_i	Upper bound of PI generated for sample i
$v_n(t)$	Velocity of particle n at step t of PSO
w	Inertia weight for velocity update during optimization with PSO
$x_n(t)$	Position of particle n at step t of PSO
$\alpha^{(t)}$	Temperature ratio between step t and step $t-1$ during annealing with SAMURAI
β	SAMURAI meta-parameter that controls interface of hybrid algorithm SAMURAI-Adam
γ	Binary parameter used for CWC calculation
δ	SAMURAI meta-parameter that controls granularity of search
S_t	The purchased energy from the grid at time t

F. Catthoor et al. (eds.), *Energy Production, Load and Battery Management Framework with Supporting Methods for Smart Microgrids*,
https://doi.org/10.1007/978-3-031-92025-7

G_t	PV energy production at time t
E_t	The charging/discharging energy of the battery system at time t
D_t^{base}	Energy consumption of the non-shiftable appliances at time t
D_t^{shft}	Energy consumption of smart appliances at time t
Y_t	Energy wasted inside the system at time t
$D_{m,p}^{app}$	The consumed energy of appliance m in phase p
C_{nom}	Normal battery capacity
C_{cell}	Battery cell capacity
N_p	The number of parallel-connected cells
N_s	The number of serial-connected cells
V_{OC}	Battery open-circuit voltage
V_{batt}	Battery terminal voltage
I_{batt}	Battery current
E	Battery charging/discharging energy
R_{total}	Total aggregated battery resistance
T_L	Appliance schedule planning period
δt_L	Time resolution
ξ_t	The electricity price from the power company at time t
E^{ref}	The reference values for battery energy
ϵ	The acceptable error from E^{ref}
α_t	Data collected from sensors for HVAC system.
s_t	Control actions (thermostat set-points) at time-step t
$E(s_t, \alpha_t)$	Energy consumption at time-step t
$C(s_t, \alpha_t)$	Thermal comfort (PPD) at time-step t
ω	Relevant importance of energy cost or thermal comfort improvement.
T_t^{out}	Outdoor temperature at time-step t
R_t	Solar radiation at time-step t
T_t^{in}	Thermal Zone Indoor temperature at time-step t
H_t	Thermal Zone Indoor Humidity at time-step t
C_{est}	Estimated thermal comfort
C_{real}	Actual thermal comfort
E_{est}	Estimated energy consumption
E_{real}	Actual energy consumption

Index

F. Catthoor et al. (eds.), *Energy Production, Load and Battery Management Framework with Supporting Methods for Smart Microgrids*,
https://doi.org/10.1007/978-3-031-92025-7

MIX
Papier aus verantwortungsvollen Quellen
Paper from responsible sources
FSC® C105338

If you have any concerns about our products,
you can contact us on
ProductSafety@springernature.com

In case Publisher is established outside the EU,
the EU authorized representative is:
Springer Nature Customer Service Center GmbH
Europaplatz 3, 69115 Heidelberg, Germany

Printed by Libri Plureos GmbH
in Hamburg, Germany